THE Hillier GARDENER'S GUIDES

The Winter Garden

Jane Sterndale-Bennett

A DAVID & CHARLES BOOK

David & Charles is an F+W Publications Inc. company
4700 East Galbraith Road
Cincinnati, OH 45236

First published in the UK in 2006

A catalogue record for this book is available from the British Library.

ISBN-13: 978-0-7153-2538-4 hardback
ISBN-10: 0-7153-2538-8 hardback

ISBN-13: 978-0-7153-2304-5 paperback
ISBN-10: 0-7153-2304-0 paperback

Printed in Singapore by KHL Printing Co Pte Ltd
for David & Charles
Brunel House Newton Abbot Devon

Produced for David & Charles by
OutHouse Publishing Winchester, Hampshire SO22 5DS

Series Consultant Andrew McIndoe

For OutHouse Publishing:
Series Editor Sue Gordon
Art Editor Robin Whitecross
Editor Lesley Riley
Design Assistant Caroline Wollen
Proofreader Audrey Horne
Indexer June Wilkins

For David & Charles:
Commissioning Editor Mic Cady
Designer Sarah Clark
Production Controller Beverley Richardson

Page 1 from left to right: *Mahonia* × *media* 'Winter Sun', *Galanthus, Cornus sericea* 'Flaviramea'.

This page: *Ilex aquifolium* 'Madame Briot', *Vinca minor* 'Illumination', *Heuchera*, *Prunus incisa* 'Praecox', *Helleborus* × *hybridus*, *Betula albosinensis* 'Bowling Green'.

ORNAMENTAL PLANT OR PERNICIOUS WEED?

In certain circumstances ornamental garden plants can be undesirable when introduced into natural habitats, either because they compete with native flora, or because they act as hosts to fungal and insect pests. Plants that are popular in one part of the world may be considered undesirable in another. Horticulturists have learned to be wary of the effect that cultivated plants may have on native habitats and, as a rule, any plant likely to be a problem in a particular area if it escapes from cultivation is restricted and therefore is not offered for sale.

Visit our website at www.davidandcharles.co.uk

David & Charles books are available from all good bookshops; alternatively you can contact our Orderline on 0870 9908222 or write to us at FREEPOST EX2 110, D&C Direct, Newton Abbot, TQ12 4ZZ (no stamp required UK only); US customers call 800-289-0963 and Canadian customers call 800-840-5220.

Contents

INTRODUCTION 4

INTRODUCING THE WINTER GARDEN 6
History of winter gardens 8
Winter gardens to visit 10
Through the window 12
In the house 14
Winter tasks 16
Speaking botanically 20

ALL-YEAR WINTER GARDEN 22
Lingering autumn 24
Approaching spring 32
Planning for all seasons 34
Climate change 40

EVERGREENS 42
Classic evergreens 44
Conifers 50
Architectural plants 52
Persistent perennials 54

WINTER FLOWERS 62
Bulbs 64
Flowering perennials 72
Flowering trees and shrubs 80

WINTER FOLIAGE 86
Green 88
White, cream and silver variegated 92
Evergold 96
Gold variegated 98
Brown, tan and orange 100
Red, purple and black 102
Eversilver and blue 104

SITUATIONS 108
Clay soil 110
Acid soil 112
Chalky soil 116
Sandy soil 118
Wet conditions 120
Woodland 122
Walls and fences 124
Pots and containers 128

THE SENSORY GARDEN 132
Light 134
Scent 136
Texture 144
Movement 150
Taste 152

AUTHOR'S CHOICE:
favourite winter planting groups 154

Index 156

Introduction

I have been selecting plants to grow in my garden in north Hampshire for the past 25 years and over this time I have become more and more convinced that the garden should be enticing and inspiring all year round, and particularly in the winter months. However, I do not have a great deal of space and so cannot set aside special areas for seasonal plantings; my plants have to grow in close harmony with each other and perform in several seasons. In these pages I aim to show how to achieve a garden that will be a delight in winter and continue to please the eye in summer as well.

Over many years of gardening I have come really to appreciate the value of good foliage. Flowers may come and go, but the leaves, in all their myriad colours, have a much longer life. Entrancing combinations can be made to carry the garden through the seasons, and by taking into consideration the texture of bark and stems, the intriguing shapes of dried grasses and the satisfying outlines of evergreen shrubs, you will be tempted outside to enjoy your winter garden whatever the weather.

In *The Winter Garden* I have included examples of all-year flowerbeds, suggested plants that remain interesting for a very long time, and addressed some of the many planting situations a gardener might meet. I have paid close attention to treasured winter flowers, many providing enticing scent as well as attractive blossoms, and covered the various leaf colours in detail.

Although I have concentrated on the winter months, I have also included the autumn plants that are lingering ever later as well as many early-arriving spring plants. With the changes in climate that seem to be under way, the seasons are beginning to merge – although I hope we still have the snowy scenes and crisp frosts that so characterize winter.

I hope that this book will inspire you to try out fresh ideas in your own garden, using both plants that are completely new to you and long-standing favourites. Then you too, I have no doubt, will look forward to winter with keen anticipation, waiting for the first hellebore to flower and the first snowdrop to open.

Jane Sterndale-Bennett

Jane Sterndale-Bennett died on 24 December 2005, shortly after completing this book.

ETHOS OF WINTER GARDENING

In high summer the beds and borders cry out to be noticed, lupins and poppies vie with the roses for vibrant colour, and the scents from flowering shrubs hang heavy in the air. Even in spring drifts of daffodils dazzle the eye and in autumn the bonfire of colouring leaves cannot fail to impress. All is much quieter in winter. You have to search out the few flowers that bravely defy the weather, bend down to admire the little yellow aconites with their ruffs of green, lift up the hellebore flowers to see the intricate pattern within, and examine the wispy flowers of the witch hazel at close quarters then stand back to pick up the scent. Look at the structure of leafless shrubs, admire the rich colouring of some stems and delight in the tiny splashes of green of the awakening leaf buds.

On a walk round the garden you will wonder where the waft of unexpected scent is coming from – the small ivory flowers of the unobtrusive Christmas box might be the source – and enjoy the excitement of the first snowdrop buds piercing through the soil. From inside the house notice how the sun shines on the architectural foliage of the fatsia or the polished mahogany of bergenia leaves and marvel at the powdering of snowy-white or palest pink petals on the winter-flowering cherry. The winter garden rewards close inspection and those who garden only for the summer season are missing some of nature's pleasures.

The RHS Award of Garden Merit 🏆

Many of the plants in this book have a symbol after their names. This denotes that they have been awarded the Royal Horticultural Society's Award of Garden Merit (AGM). To qualify for the award, a plant has to have more than one point of interest as well as a unique property. It has to be easy to grow, of sound constitution and, most importantly, a good all-round garden plant. This is not to say that a plant without an AGM is not worth growing. It means that, to date, the Royal Horticultural Society has not trialled it or the plant does not meet all these criteria.

LEFT: *Miscanthus* after a snowfall.

INTRODUCING THE WINTER GARDEN

The garden need not be forgotten in winter and simply left to await the first shoots of spring. Some plants are at their best at this time of year, whether for flowers or for foliage – so by planting with winter in mind there will be a whole new season to savour, viewed through the window or on a walk around the garden. Delicate flowers and colourful stems can be cut and brought indoors, work can be undertaken in preparation for the year ahead, and a visit to one of the growing number of gardens open to the public in winter will provide fresh ideas for planting schemes.

RIGHT: The author's garden at White Windows, Longparish, Hampshire.

History of winter gardens

For centuries gardeners were appreciative of plants and design schemes that cheered up the winter months, but such winter-flowering plants as there were had to be tough and hardy, and mostly came from northern temperate climates. Then, in the early 19th century, came glasshouses, and soon wealthy landowners were vying with each other to fill their new conservatories with exotic plants that flowered in the winter. Camellias were first introduced in this way; originally suffocated in overheated greenhouses, they eventually found their way into cooler orangeries and from there into sheltered areas of the garden. Even cottage gardeners, lacking the means to provide winter protection, found they could tuck in special treasures; many porches were to be seen dripping with yellow jasmine, and Christmas roses flourished by the back door.

Left: The magnificent domed, early Victorian Palm House in Belfast's Botanic Gardens. When exotic shrubs like camellias (above, *Camellia japonica* 'Nobilissima') were first cultivated in Britain, they were grown in glasshouses.

THE PUBLIC WINTER GARDEN

During the Victorian era many municipal authorities built huge ornamental conservatories as entertainment venues, and the term winter garden soon began to be applied to these people's palaces. Used for band concerts, dancing and other forms of public entertainment, they were filled with plants of exotic form and texture, often richly scented. Many cities built these structures: Brighton, Bournemouth, Edinburgh, Glasgow and Belfast all had fine winter gardens. The most famous was the Crystal Palace, which was moved to Sydenham, in south London, in 1854 after being constructed for the Great Exhibition in Hyde Park in 1851. It housed a winter garden, with plants and trees from every climatic zone filling the transept and aisles along the nave. It was destroyed by fire in 1936.

These winter palaces fell from favour after World War I as they became too expensive to heat and maintain. However, modern technology has enabled a come-back. The city of Sheffield, for example, bravely built a brand new Winter Garden glasshouse, which opened to great acclaim in 2002.

In our smaller, modern gardens many of us do not have space to set aside an area purely for winter plants. Botanical and public gardens have more scope, however, and many are planting special winter borders or creating entire winter gardens – no longer enclosed in glass, and part of the general garden landscape. Designed for winter viewing, these are becoming increasingly popular as winters get milder and more and more interesting plants can be grown. The Cambridge University Botanic Garden was one of the first to establish a separate winter garden, in 1979.

The all-year-round garden, consisting of conifers and heathers, was popular in the 1980s. Above: Adrian Bloom's garden, Foggy Bottom, in 1989.

THE WINTER GARDEN AT HOME

The way we use our private gardens changed dramatically during the 20th century. After World War II the interest in ornamental plants was rekindled, and we started to grow more flowers, in addition to fruit and vegetables. The garden had traditionally focused on summer, with the result that colourful bedding displays and roses all too often left a rather bleak and barren garden in winter. But as homes became more comfortable and windows larger, and as gardening began to get more exposure in print and broadcast media, so interest in the garden as an extension of the home began to grow, and the look of the garden in winter became more important. The range of garden plants was expanding: the legacy of the plant hunters, a vast array of ornamental plants mainly introduced from the mid-19th century onwards, only started to make a real impact in private gardens in the second half of the 20th century.

In the 1970s and 1980s winter gardens appeared to consist of ornamental conifers planted amid carpets of flowering heathers. Attractive as these seemed at the time, they were very static, and unchanging in the summer months, and after a decade or so interest began to wane. Today we still use conifers and heathers, but in combination with a range of deciduous shrubs, perennials and bulbs, and to much better effect. Winter-flowering plants and evergreens can easily be incorporated within borders, while many plants for winter and year-round interest thrive in pots. We have become more aware of the shapes and textures of plants, the colours of stems and bark, and the importance of structure and height in the garden.

Front gardens are particularly well suited to winter plantings. Plants can be observed at close quarters, and scents appreciated, by everyone coming and going from the house. The house wall often provides extra protection for a tender shrub, and the eagerly awaited little winter bulbs can be enjoyed just as soon as they emerge.

THE FORMAL GARDEN

Ever since the first 'paradise' gardens in Persia, walls and hedges have been used to define the garden and to separate it from the surrounding landscape. In the Middle Ages the monks cultivated their herbs and vegetables in walled gardens, and in the grand houses of the 18th and 19th centuries the kitchen garden was often a walled enclosure, distant from the house. Elizabethan gardeners, with a very limited choice of plants, created formal parterres and knot gardens, which were often filled with coloured gravels rather than plants.

All of these gardens had structure, which provided a pleasant picture throughout the year. Today formal gardens with clipped hedges and topiary, sometimes with a reflecting pool and statuary, provide a satisfying, static picture in winter. While they are pleasing in their symmetry, particularly when enhanced by frost or a blanket of snow, they do rely on immaculate housekeeping.

Winter gardens to visit

A number of gardens have established substantial collections of plants for winter interest and are well worth visiting for ideas and inspiration, and to see what a colourful season this can be. Others are famous for their drifts of snowdrops, which give pleasure both to connoisseurs and to the many visitors who come simply to delight in the first signs of spring. Formal gardens with parterres, topiary and structure are magical in winter light and are fine places for frosty walks. Increasingly popular for winter visits are the big glasshouses, which hark back to the early Victorian people's palaces.

The Winter Garden at the Sir Harold Hillier Gardens, the largest outdoor winter garden in Europe.

The **Sir Harold Hillier Gardens** (formerly known as the Hillier Arboretum), near Romsey, Hampshire, cover more than 160 acres and house one of the largest collections of woody plants in the world. Started in the 1950s, they are the work of the late Sir Harold Hillier and are now managed by Hampshire County Council. An excellent place to see trees and enjoy winter bark and stems, they also boast the National Collection of *Hamamelis* (witch hazels). The Winter Garden, established in 1996 and located close to the visitor centre, brings together plants of all types that excel in the winter months, giving an idea of just how good a garden of this type can look in any season.

The magnificent National Trust garden of **Anglesey Abbey**, near Cambridge, offers plenty of winter interest. More than 100 varieties of snowdrop can be found, together with a pinetum and a mile-long Winter Walk with trees and shrubs chosen for their bark, catkins or coloured stems. The site of the Winter Garden in **Cambridge University Botanic Garden** was landscaped to form a shallow valley that is flooded by the light of the afternoon sun. In winter this illuminates drifts of dogwood and willow stems set among a wealth of plants grown for winter flowers and foliage.

SNOWDROPS

There are many gardens to visit in the snowdrop season. In some you can admire great drifts of naturalized *Galanthus nivalis*; in others you can compare a host of rare varieties, all with their individual peculiarities.

Good gardens to see snowdrops include: Audley End, Essex; Benington Lordship, Hertfordshire; Cambo Gardens, Fife; Chirk Castle, Clwyd; East Lambrook Manor, Somerset; Fountains Abbey, North Yorkshire; Heale Garden, Wiltshire; Hodsock Priory, Nottinghamshire; Lacock Abbey, Wiltshire; Painswick Rococo Garden, Gloucestershire.

GALANTHOPHILIA

True galanthophiles visit gardens with specialist snowdrop collections such as Colesbourne Park in Gloucestershire. Other smaller gardens, like Brandy Mount, in Hampshire, are open through the National Garden Scheme on selected dates in late winter. Here visitors can compare the finer points of shape and form and the unique and delicate markings on snowdrop flowers.

Snowdrops – harbingers of spring.

Top: The Winter Garden at RHS Garden Rosemoor. Above: *Cyclamen coum* at RHS Garden Wisley.

Ham House, Surrey, is a 17th-century house on the banks of the Thames with a garden famous for its parterres of yew, box and lavender punctuated with statuary and urns.

The **Royal Botanic Garden Edinburgh** has a quite superb collection of trees and plants on a fine hillside site and, with 450 types of conifer and a fine Victorian palmhouse, is as popular a place for a walk in winter as in any other season.

The **Royal Botanic Gardens, Kew**, in London, boast a number of splendid temperate and tropical glasshouses holding mature specimens of plants from all over the world. The orchids attract many visitors in mid- to late winter.

RHS Garden Harlow Carr, Harrogate, North Yorkshire, a beautifully tranquil garden, laid out a winter walk in 2006.

RHS Garden Rosemoor, Great Torrington, Devon, was opened by the Royal Horticultural Society in 1990. The Winter Garden is one of the themed areas, featuring plants for foliage and texture as well as for flowers and fragrance.

RHS Garden Wisley, Woking, Surrey, the principal garden of the Royal Horticultural Society, excels in winter with its pinetum and heather gardens, willow and dogwood stems by the water, witch hazels on Battleston Hill and in Seven Acres, and an abundance of snowdrops.

The **Savill Garden**, Surrey, is a stunning woodland garden on acid soil: birches, alders, rhododendrons, witch hazels and snakebark maples all excel in winter, and the beds near the visitor centre display brilliant colour.

Sheffield Winter Garden is one of the largest temperate glasshouses to be built in recent years, home to 2,000 plants from around the world, including tree ferns and palms.

Threave, Dumfries and Galloway, the National Trust for Scotland's gardening school, has a fine display of colourful stems, hollies, winter-flowering shrubs and heathers.

Through the window

The view from the windows of the house keeps you in touch with your garden at all times of the year, but especially so in the winter months, often providing the only picture of the garden on inclement days. Careful placing of plants and objects, both large and small, results in a changing scene as winter advances, and sometimes a view from an upstairs window will show aspects of the garden not seen from ground level. In late autumn, as deciduous trees shed the last of their leaves, distant views reappear and the bare bones of the garden are exposed.

Winter light on shining evergreens and the russet leaves of beech and parchment grasses creates an ever-changing picture from the windows of the house.

Evergreen and deciduous shrubs and trees are important elements of the garden throughout the year, providing the backbone of the planting in summer and the essential fabric of the garden in winter. A Himalayan birch with a gleaming white trunk, such as *Betula utilis* var. *jacquemontii*♀, can be used as a focal point, drawing the eye into the depths of the garden, as can the narrow spire of an Irish yew (*Taxus baccata* 'Fastigiata'♀).

Walls and fences clothed in deciduous climbers come to the fore as the leaves fall and the bare surface reappears. Strongly coloured wooden fences can be obtrusive, but those treated with a naturally coloured preservative sit more easily with the planting. A wall or fence at the back of a border suddenly becomes prominent in winter, even if viewed through deciduous shrubs. Architectural evergreens like × *Fatshedera lizei*♀, with large, leathery, palmate leaves, can bring interest here, as can the many forms of variegated ivy, especially when their outlines are etched by frost.

Hedges composed of evergreens provide solid structure so valuable in the winter months, but deciduous plants can also be included: with beech or hornbeam, for example, the drying leaves remain on the branches, and these give an excellent effect in winter as they fade to a warm chestnut brown, a pleasing colour on a grey day.

Garden ornaments come in many forms and in winter they help to enliven the scene. Elegant plant supports that have held up clematis or other climbers can be cleared of their summer growth and then themselves become focal points; metal arches too will define a distant view. A large stone urn dripping with pelargoniums or fuchsias in summer is a very pleasing shape when empty in winter. Statues bring form to a garden and, if placed where the sun can strike them, will draw the eye to a quiet corner. All gardens should have seats and, if well chosen, these can be attractive as well as useful. Even mundane objects can have an ornamental value: think of a collection of galvanized watering cans or mellow clay rhubarb pots.

Nearer the house winter-flowering shrubs and perennials provide splashes of colour. A number of evergreens have silver or purple leaves and some have foliage variegated in cream or gold; desiccated grasses bring in tones of straw and buff, and the stems on many shrubs are a rich chestnut brown or even orange, while perennials with overwintering leaves come in shades of bronze, claret and purple. All these colours dispel the myth that the garden in winter is dull and drab and best ignored.

In contrast, a formal garden with well-clipped topiary presents a satisfying picture in monochrome, often highlighted by a fall of snow. Neat box hedging, a rectangular pool reflecting the low winter light and a well-placed statue in front of a precision-trimmed yew hedge need little attention over the winter and are often perfectly framed by the windows of the house.

THE LAWN IN WINTER

The lawn becomes more prominent in the winter garden. In summer, when green predominates in beds and borders, the lawn edges are blurred and blend into the planting. In winter the lawn becomes a well-defined green form and its proportions in relation to the rest of the garden become more apparent. This is an excellent time to evaluate the lines and shape of the lawn to ensure a pleasing picture throughout the year.

The bold foliage of a bergenia and a garden bench provide a strong focal point in this picture seen through the house window.

One of the pleasures of winter is the arrival of the first flowers. The little winter bulbs – pristine white snowdrops, sugar-pink cyclamen and canary-yellow aconites – should be planted close to the house where you can see them without venturing outside. These will be followed by an early Christmas rose (*Helleborus niger*♡), with delightful cups of cream filled with yellow stamens, and then the rich purples, reds and pinks of the Lenten rose (*Helleborus* × *hybridus*). At a higher level the winter-flowering shrubs delight the eye: pastel pink viburnums, vivid yellow jasmine (*Jasminum nudiflorum*♡) and the distant gleam of a yellow witch hazel (*Hamamelis*) caught by the watery sun.

A collection of pots in differing sizes, grouped together close to the house, keeps the interest going. Pots of clipped box and the invaluable *Skimmia japonica* 'Rubella'♡, with its glossy, dark green leaves and incredibly long-lasting ruby buds, make a good base. Then add in small pots of winter-flowering plants in succession – snowdrops, aconites, cyclamen and pansies. The changing picture will entertain all winter, and before long the skimmia will be showing white scented flowers to go with spring bulbs.

In the house

The garden can be brought indoors in winter with pots of snowdrops, aconites and primroses placed on window sills or tables where they can be inspected closely. Bowls of hellebore flowers, floating in water or sitting on moss, form a stunning centrepiece, just a few sprigs from a scented shrub will perfume a room, and colourful branches make a pleasing winter decoration.

A few muscari dug from the garden come into bloom when brought indoors.

INDOOR POTS

Snowdrops deserve close study and if a small clump is dug up from the garden in tight bud, potted up and brought into the house, you will be able to see how the initially upward-pointing buds gradually bend over and the now hanging petals spring apart to reveal the green markings within. A fairly deep container will be needed and some moss or dried leaves to hide the bare stems. When the flowers have gone over, the clump can be divided and planted outside again. Other bulbs such as crocus, muscari and fritillaria can be treated in a similar way, as can winter aconites and early primroses. Towards the end of winter it is often possible to find a few precocious wild violets tucked away in a corner of the garden, and their white, pink or violet flowers make a charming addition to a display.

CUT FLOWERS

Winter flowers are rather scarce in the garden and so only a very few can be spared for indoor decoration. Small bunches of these precious flowers look best in little vases or in shallow bowls filled with moss. Hellebore flowers do not last in water on their stems but are lovely when displayed floating in water in a glass bowl or in a moss-filled dish (see page 74); this treatment provides the perfect opportunity to study the wonderful variations of spots and staining on the inside of the flower cups.

MISTLETOE

Mistletoe (*Viscum album*) has long been part of Christmas greenery, believed to bring good luck and fertility to the household, as well as protection from evil. It is a semi-parasitic plant that sustains itself on the branches of mature trees, usually apples, but also pears, hawthorns, limes and others. Because of its parasitic nature, it was always given special significance by the ancients and featured in their religious ceremonies. The Druids considered it had miraculous powers and used ceremonial golden sickles to cut it from the trees at the winter solstice. The Romans thought it brought peace, and in northern Europe it was revered by the Celts and Goths. Its strong pagan associations led it to be banned from churches; instead it was assigned to kitchens and hung up over the door where 'whatever female chanced to stand under it, the young man present had a right of saluting her and of plucking off a berry at each kiss'. Even today we hang up small bunches in entrance ways and kisses can be claimed. Recently extracts of the plant have been used medicinally.

Often a few sprigs of a scented shrub like a daphne, a sarcococca or a shrubby honeysuckle are all that is needed to fill a room with perfume. Witch hazel picked at the brown bud stage will slowly open, releasing its distinctive scent; *Clematis cirrhosa* var. *balearica* has a delicious lemon fragrance, and *Viburnum farreri*🏆 smells of almonds.

Many spring-flowering shrubs can be persuaded to flower early if picked at bud stage and brought indoors. *Forsythia ovata* 'Tetragold' flowers earlier than most, and cherry and almond blossom will also respond to this treatment.

The twisted stems of the corkscrew hazel, *Corylus avellana* 'Contorta', make a stunning decoration in this elegant vase.

Picked stems of snowdrops look best in little containers with a few leaves – the marbled leaves of *Arum italicum* subsp. *italicum* 'Marmoratum'🏆 are good companions. The Algerian *Iris unguicularis*🏆 is best admired indoors: a few tight buds will unfurl to reveal intricate patterns of lilac and yellow on the petals and give off a subtle perfume.

STEMS

The colourful stems of dogwoods make a good indoor decoration. A tall, narrow-necked porcelain jar filled with wands in red, yellow, purple and green will be long-lasting; in fact, some of the stems may produce roots and so provide more plants for free. Willows with coloured stems can also be used and if ones that bear early catkins are chosen, even better.

The corkscrew hazel, *Corylus avellana* 'Contorta', also known as Harry Lauder's walking stick, has curiously contorted branches from which fat yellow catkins dangle in late winter, and the twisted willow, *Salix babylonica* var. *pekinensis* 'Tortuosa'🏆, has thin writhing branches; these can be arranged by themselves or mixed with trailing stems of ivy and evergreen leaves.

WREATHS

It is a growing custom to place wreaths on front doors and gates over the Christmas period. Most garden centres and many markets sell wreaths of evergreen leaves such as holly or conifer, which can be enhanced with variegated foliage, berries, fir cones and even old man's beard, *Clematis vitalba*. Contemporary wreaths can be created by weaving stems of coloured dogwoods and willows into circles.

Holly berries are invariably taken by birds just before you go outside to pick them yourself, but the fat berries on the female skimmias are often ignored until the end of winter, and these look just as well in wreaths. Rose hips too are very effective and many shrub roses produce good-sized hips. Strip the stems of their leaves and tuck the berries into the wreath. A few of the ruby-red buds on *Skimmia japonica* 'Rubella'🏆 and ivy flowerheads will add interest and texture, and fruit such as small apples, oranges, lemons and limes can also be used.

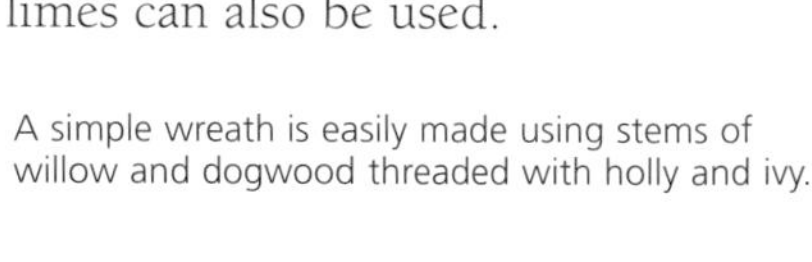

A simple wreath is easily made using stems of willow and dogwood threaded with holly and ivy.

Winter tasks

For gardeners not content to sit by the fire and dream about summer glory there are plenty of tasks to carry out in the winter. Paths, patios and lawns need to be cleared of dead leaves, deciduous hedges may be cut, climbers need to be tidied up and late-flowering shrubs should be pruned. Weeds seem to grow regardless of temperature and season, and digging and preparation for the coming year will get the garden in good shape for the spring and summer.

The dried flower stems of grasses and perennials such as phlomis continue to provide interest and food for hungry birds if left on the plants until late winter.

COLLECTING LEAVES

Collecting fallen leaves is just as much a winter task as an autumn one since many deciduous trees hang on to their foliage well into winter. Leaves should be removed from the lawn as soon as possible, since wet leaves lying on the grass keep out the light and can cause damage. If the weather is reasonably dry the easiest way to collect them is with a rotary lawn mower: put it on a high setting and simply mow them up. This has the advantage of chopping the leaves and incorporating a few grass cuttings, which will help with the composting process.

Leaves take a long time to break down; a good way to compost them is to put them into large black plastic sacks. Push the leaves down well and if they are dry dampen them before adding more. Tie up the top of each sack and make a few holes in the bottom with a garden fork. Store the sacks in a shaded corner out of the way and the leaves will break down into the most wonderful leaf mould after a year or so, depending on the type of leaf. Oak leaves take the longest; they are notoriously tough and resilient.

DIGGING

It is well worth making the effort to dig over any empty beds and the vegetable plot in early winter, especially on clay soils. Do not break the soil down into a fine tilth after digging; instead leave it rough and 'slabby'. This exposes the soil to frost action over the winter months, which will help it to break down naturally. These beds can be manured in early winter; any excess nitrogen in the manure will be washed away by winter rain.

Well-rotted compost applied to beds and borders in winter will suppress weeds and will be dragged down into the soil by earthworms.

CLEARING AND MULCHING

Traditionally herbaceous borders were cleared in the autumn, the plants lifted, divided and replanted and heavy amounts of manure dug into the soil. Apart from being very hard work this left the borders bare in winter. It is now much more acceptable to leave many perennials and grasses uncut in the autumn in order to enjoy the seedheads and the attractive colours of the faded leaves. In late winter they should finally be cut back to the ground to make way for new growth. At the same time apply a good layer of garden compost, spreading it over the surface carefully to avoid smothering the crowns of resting herbaceous plants; do not carry out this work if the ground is frozen or the frost will be trapped in the soil.

In mature mixed borders and between shrubs it is virtually impossible to cultivate the soil and incorporate organic material such as compost. If the compost is spread over the soil surface in early winter, smothering any fallen leaves that have collected on the beds, earthworms should do most of the cultivation for you as they drag compost and leaves down into the soil. Over time this will build up the humus content of the soil, whatever its type.

DIVIDING AND PLANTING

Given the milder, wetter winters of recent years, it is probably best to leave the task of dividing herbaceous perennials until early spring when they are ready to come into growth; autumn divisions can rot in wet conditions if they have not made enough root growth. The same is true of planting container-grown perennials if you garden on heavy soil.

Container grown trees and shrubs can be planted at any time, although planting in autumn and early winter allows the roots to establish in the soil before growth starts the following spring. Bare-root plants can be planted during

Virtually all soils are improved by the addition of organic matter. Use winter to collect leaves, prunings and general garden clippings to make plenty of rich compost.

PONDS

Fish keep a low profile in the winter months, often disappearing to the bottom among the weeds. Regularly remove any fallen leaves from the surface with a net. Also remove any stems and leaves of marginal plants that have collapsed into the pool. This work should be carried out until midwinter, but if frogs use the pond to spawn in, all should then be left well alone: if the weather is mild, spawning can start in late winter.

The pond at the Sir Harold Hillier Gardens, Hampshire.

any mild spell from early winter until early spring. Trees should be securely staked and tied, and new shrubs may need staking on exposed sites: newly planted subjects that rock in the wind will never establish themselves.

BORDER DESIGN

Winter is an excellent time to evaluate the design of the garden and to consider altering the shape of beds and borders without hindrance from the clutter of the summer foliage. If radically changing the shape of a bed or creating a new path, it is often best to lay out a hosepipe or to cut out a narrow channel in the lawn with an edging moon in the shape you want and live with it for a while before committing spade to turf. Make a note in summer of any shrubs that need curbing or removing entirely and carry out this work in winter.

WINTER PROTECTION

Some more tender shrubs and perennials will benefit from a covering of horticultural fleece in frosty weather. This should be loosely draped over the plants and secured near the base to prevent it blowing away. There are various grades of fleece: the slightly heavier, thicker ones give better protection. Pay particular attention to the less hardy evergreens such as cordylines, phormiums, ceanothus and large-leaved hebes. Exotics such as bananas and tree ferns need to be packed round with cut bracken fronds or straw in a netting container. Plants in pots can be vulnerable, especially at the roots. Pots can be lagged with straw or fleece or wrapped in layers of hessian sacking. Moving pots close to the walls of the house where they will be sheltered by the eaves affords excellent protection. None of the materials used for protection looks beautiful in the garden; however, fleece or sacking need only be in position through the worst of the cold weather.

PLANT LABELS

Plant labels in a private garden can be a vexatious problem, particularly in winter when a scattering of white sticks or coloured tags looks inappropriate – but plants do need their names. One solution is to use black labels with writing in a fade-proof, fine-point silver marker pen. These are unobtrusive and can be tucked beside the plant and will last for many years.

WINTER PRUNING

TREES Winter shaping of deciduous trees may improve their appearance in summer. If trees are grown in mixed borders, it is advisable to prune out the lower branches to lift the canopy so that nearby plants get enough light. Also cut out any crossing branches in the centre of the tree and thin the crown, which will allow more light and rainfall to reach the plants below. This work is best carried out in midwinter.

SHRUBS Shrubs that flower on new season's growth need to be hard pruned in late winter, in order to produce strong new shoots and plenty of flowers later on. Most buddleias should have this treatment, as should caryopteris, ceratostigma and perovskia; the *Hydrangea paniculata* cultivars should also be pruned in this way, otherwise they produce light spindly growth instead of vigorous arching branches. Stems should be cut back by two thirds or to the lowest pair of buds. The small-flowered hardy fuchsias can be cut right back to ground level.

With dogwoods (*Cornus*) and willows (*Salix*) grown for their coloured stems, cut back roughly half of the stems to ground level in late winter or early spring; this encourages vigorous new shoots for colour next winter. For elders (*Sambucus*), hard prune young plants to promote vigorous growth.

The growth of some shrubs, such as potentillas and spiraeas, can be controlled by cutting them hard back every three years; in other years they should be given a light trim to remove old flowerheads and the untidy growth from the outside of the bush.

Evergreens such as elaeagnus and photinia can be pruned in late winter to control size and shape before the new growth flushes through in early spring. Avoid clipping to an unnatural shape unless the plant is being formally trained. Cut out stems that are encroaching on other plants or spoiling the shape of the bush.

Roses can be pruned in midwinter. Old-fashioned shrub roses and English roses are pruned lightly to encourage plenty of side-branches at the tops of the stems: these produce the flowers. Hard pruning promotes vigorous growth and should be confined to hybrid tea and floribunda roses; these can be cut back to a few buds above ground level. In all cases remove dead and diseased wood and prune to outward-facing buds to promote an open shape.

CLIMBERS The late-flowering *Clematis viticella* cultivars need to be cut to within 30cm (12in) of the ground and all the spent growth pulled away. Vines such as *Vitis coignetiae*♀ and *Vitis vinifera* 'Purpurea'♀ should be pruned in early winter – any later and they will bleed (lose sap), possibly fatally. Cut back the light side-branches to three buds from the main stems.

Wisterias need pruning twice each year. Any excess growth is removed in midsummer then, in winter, all long, thin trailing shoots are cut back to a few fat buds from the main stem; this will encourage the production of small side-shoots (spurs) on which the flowers will be borne.

Climbing and rambling roses need to be checked and tied in to their supports. Take young flexible growths and arch them along a fence or wall; also cut out some of the older growth to make way for vigorous new shoots. Climbing roses will flower much more prolifically if they are trained in this way.

SHRUBS TO LEAVE ALONE IN WINTER

Any shrubs that flower in spring and early summer should be left alone in winter. Pruning will cut off the wood that will produce the following season's flowers. These include: syringa, philadelphus, forsythia, deutzia, weigela and ribes.

Above: Both buddleia (top) and perovskia (centre) should be hard pruned in late winter, cutting back stems by at least two thirds. Below: Cornus grown for their coloured stems should have at least half the stems cut back to the ground in late winter.

Speaking botanically

Latin names, as opposed to common names, may be confusing at first, but they are essential when it comes to accurate communication. A plant's common name may vary from one area to another, with local names being used in different countries for the same plant. Botanical Latin is an international language that can be understood by gardeners from all over the world and it often conveys constructive information about a plant.

The botanical name *Galanthus* identifies these delicate white and green winter beauties anywhere in the world.

The snowdrop provides us with an excellent example of why Latin botanical names are so useful. In Britain the plant is known as a snowdrop, in Germany it is *Schneeglöckchen* (snow-bell) and in France it is *perce-neige* (snow-piercer) – and there are doubtless many more names as the bulb is widely distributed throughout Europe and the Middle East. By using the botanical name, *Galanthus* (derived from the Greek *gala,* milk, and *anthos*, flower), confusion is avoided.

GENERA AND SPECIES The binomial (two-name) system of naming plants was first set out by Linnaeus in his book *Species Plantarum* (1753). The genus comes first, always with an initial capital letter, followed by the species name in lower case; both are usually written in italics. The genus name describes the group to which the plant belongs; although there can be much variation within a group, all will share some characteristics. All *Galanthus* are small bulbs but while most flower in the winter, some flower in late autumn. The species name is used to differentiate one from another, so *Galanthus nivalis*, the common snowdrop, which flowers in winter, is one species, while *Galanthus reginae-olgae*, which flowers in autumn, is another.

The species name is often helpful in describing the plant as it may refer to its origins or its native habitat; for example, *Galanthus nivalis* in Latin means 'grows near snow' and *Galanthus ikariae* indicates that the plant comes from the Greek island of Ikaria. The species name may describe the flower, as in *Galanthus gracilis* (graceful), or the shape of the leaves, as in *Galanthus platyphyllus* (broad leaves). It may also refer to a botanist or plant hunter who discovered or first described the plant. *Galanthus elwesii* is named for the great traveller, sportsman and plant collector Henry John Elwes.

CULTIVARS AND HYBRIDS While the species name describes plants found in the wild, gardeners and plant breeders soon began selecting special forms of these plants so that a third name became necessary – the cultivar name, with cultivar being an abbreviation of 'cultivated variety'. The accepted way of writing a cultivar name is to place it within single quotation marks, not in italics, and with initial capital letters, as in *Galanthus nivalis* 'Viridapice' for a form with green tips to the flowers; however since 1959 it has not been permissible to Latinize new cultivar names, so now we get *Galanthus nivalis* 'Tiny'. Sometimes the parent species is unclear, in which case just the genus and the cultivar names are used, as in *Galanthus* 'S. Arnott'.

Some plants are known to be hybrids between two species and they are described by using a multiplication sign between the genus and the new hybrid name, as in *Galanthus* × *hybridus* 'Merlin', which is a cross between the two species *Galanthus elwesii* and *Galanthus plicatus*.

NAME CHANGES Changes in long-established names are a nuisance to everybody and may seem difficult to justify, but there are usually good reasons for the change. Research is taking place all the time and new information often comes to light; earlier names are discovered and modern scientists are able to detect ever more minute differences (and similarities) in the structure of plants; the species formerly known as *Galanthus caucasicus* is now believed to be a form of *Galanthus elwesii* with only one green mark on each petal instead of two.

WORDS DENOTING WINTER USED IN BOTANICAL NAMES

The chart below lists some of the many Latin adjectives denoting winter that are found in plant names. As with nouns, Latin adjectives have male, female and neuter versions and their gender generally matches that of the noun they describe. For example, *Helleborus* is a male noun, so the adjectives accompanying it are also in the male form, for instance *Helleborus niger* or *Helleborus viridis* (the female forms are *nigra, viridis* and the neuter are *nigrum, viride*). Alternative gender endings of the adjectives are given in brackets.

WINTER TERM IN LATIN	TRANSLATION	EXAMPLE
arcticus (-a, -um)	arctic	*Rubus arcticus*
chionanthus (-a, -um)	with snow-white flowers	*Primula chionantha*
frigidus (-a, -um)	growing in cold regions	*Cotoneaster frigidus*
glacialis (-is, -e)	icy cold regions especially near glaciers	*Ranunculus glacialis* *Geum glaciale album*
hiemalis or *hyemalis* (-is, -e)	of winter winter flowering	 *Eranthis hyemalis*
nivalis (-is, -e)	snow white growing near snow	*Galanthus nivalis* *Podocarpus nivalis*
niphophilus (-a, -um)	snow loving	*Eucalyptus pauciflora** subsp. *niphophila*
nudiflorus (-a, -um)	naked flowers, flowering before the leaves open	*Jasminum nudiflorum*

* trees are nearly always treated as feminine

The fascinating subject of plant names and their meanings is fully explored in Hillier's *Plant Names Explained.*

Helleborus niger with snowdrops.

ALL-YEAR WINTER GARDEN

A garden should provide interest and entertainment at any time of the year. When planning the winter garden, it is important to consider the other seasons so that new plants are coming to prominence all the time – as one plant dies down or fades into the background, another takes its place. Winter is no longer the season when all is sleeping; with the onset of climate change the autumn plants are lingering longer, spring plants are flowering earlier and more plants than ever before are flourishing throughout the winter.

RIGHT: *Sedum* 'Herbstfreude' in low winter light.

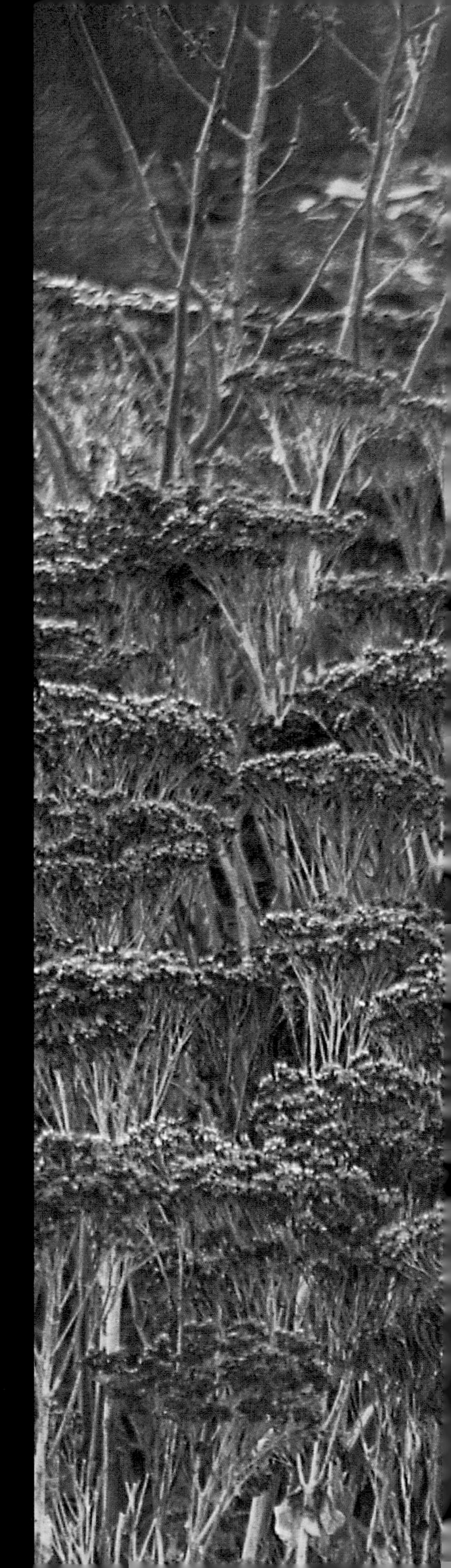

Lingering autumn

The change to winter from autumn is a slow and subtle one: a few flowers cling to some shrubs, the roses linger on, and autumn berries will remain into winter if the birds do not devour them first. The skeletons of many perennials, now mere desiccated stems and seedheads, take on a beauty of their own. The first frosts herald a colour change in many leaves and, as they fall, the structure of trees and shrubs becomes apparent and their often beautiful bark is revealed.

Many deciduous trees and shrubs retain leaves and berries well into winter, displaying warm shades against the greens and variegations of their evergreen neighbours.

Abelia × grandiflora

Hydrangea paniculata

SHRUBS AND TREES

Some shrubs flower in the autumn and then retain these flowers well into winter. ***Abelia* × *grandiflora***🏆 is a rather underrated semi-evergreen shrub, with small glossy leaves on arching stems carrying showers of trumpet-shaped, pale pink flowers. These are backed by darker, bronze-pink calyces, which persist after the petals have fallen; in fact, in a mild year the calyces remain on the plant almost into midwinter. This medium-sized shrub can be grown in the angle of a wall, where the arching stems, often tinted pink, can spread out; a brick wall that echoes the colour of the calyces is ideal. There are cultivars with variegated leaves, such as ***Abelia grandiflora* 'Francis Mason'**, and there is a hybrid, ***Abelia* 'Edward Goucher'**, which forms a smaller shrub with abundant lilac-pink flowers.

The lacy flowerheads of some **hydrangeas** fade to a pale coffee colour, adding delicate texture to the garden through winter. They can also be cut and dried and used to decorate the house.

ROSES

The last lingering roses of autumn give a great deal of pleasure and, with the milder conditions of recent years, some of the repeat-flowering roses have been producing blooms in early winter. ***Rosa* 'Felicia'**🏆, one of the hybrid musk roses produced by the Reverend Joseph Pemberton in the 1920s, is a long-lived,

Rosa virginiana

reliably repeat-flowering rose, growing to 1.2–1.5m (4–5ft), which looks especially fine arching along a picket fence. Its first flush of flowers, warm apricot-pink in bud opening to silvery pink, starts in early summer. If the trusses are cut back when the flowers fade, 'Felicia' will bloom again in early autumn, stopping only when the frosts become severe (see page 41). Of similar breeding, ***Rosa* 'Will Scarlet'**, with bright crimson flowers, usually still bears a few blooms in late autumn, and there are many modern repeat-flowering bush roses that will also keep on producing until harsh winter weather sets in. Many roses bear good hips, and the tough **rugosa roses** in particular hang on to their fat hips well into midwinter. Forming large dense shrubs, with dark green, deeply veined leaves, the rugosas are excellent growing in poor soils.

There is a rose for all four seasons in the North American ***Rosa virginiana***♡. This is a thicket-forming rose, best grown as a specimen in grass, where it will reach 1.5m (5ft) high and as much across. In summer it bears single, bright pink flowers on red stems over a long period, followed by large, bright red hips. The leaves are a fresh glossy green in spring, and in autumn they go through a kaleidoscope of colours, changing to purple, then red, orange and finally yellow. The whole bush appears on fire. When the leaves fall the young stems come to prominence, crimson in colour and carrying scarlet hips, which remain throughout winter, seemingly unappetizing to the birds. With a surrounding blanket of snow, this rose is a striking sight in winter. To encourage plenty of young growth, cut back the bush to the ground every five years or so.

A few rose blooms and leaves linger on, in mild areas often until midwinter.

Sorbus 'Joseph Rock'

Cotoneaster lacteus

Malus transitoria

Ilex aquifolium 'J.C. van Tol'

Pyracantha 'Orange Glow'

AUTUMN AND WINTER BERRIES

Many trees have colourful autumn berries that hang on into winter. The fruit on the rowans, ***Sorbus aucuparia***, is abundantly borne in some years, with the yellow berries in particular tending to be ignored by the birds until they have polished off all the red and orange ones. ***Sorbus* 'Joseph Rock'**, an excellent small tree with lovely amber-yellow berries, is a good selection. The fruit on some of the crab apples (***Malus***) remains on the branches well into winter, but often becomes shrivelled and an unattractive dark brown. ***Malus* × *zumi* 'Golden Hornet'** suffers this fate, but the delicate, small-growing ***Malus transitoria***♀ keeps its currant-sized, soft apricot-yellow fruit in good condition over a period of many weeks.

The evergreen **cotoneasters** are excellent berrying shrubs with heavy clusters of usually bright red, long-lasting fruit following a profusion of small, white to pink flowers in early summer. Some grow into large shrubs, such as ***Cotoneaster frigidus* 'Cornubia'**♀, which is semi-evergreen with arching branches laden with fruit, and ***Cotoneaster lacteus***♀, which fruits a bit later; ***Cotoneaster franchetii*** is a medium-sized evergreen with arching stems and large bright red berries (see page 110). All these are useful shrubs that thrive in almost any soil. ***Cotoneaster* 'Hybridus Pendulus'** has branches weighed down with brilliant red berries, and can be trained to grow on a single stem to form a small weeping tree. Low-growing ***Cotoneaster* × *suecicus* 'Coral Beauty'** makes a dense, arching shrub and is a better choice for ground cover than one of its parents, ***Cotoneaster dammeri***♀, whose long trailing shoots can wander too far. 'Coral Beauty' is happy in full sun and associates well with grasses and other ethereal plants, which are set off by its glossy evergreen leaves and the masses of bright coral berries.

The hollies (***Ilex***) are a large group of plants with innumerable cultivars selected for their patterned leaves as well as for their berries. ***Ilex aquifolium* 'J.C. van Tol'**♀ is a reliable choice for good bright red berries. (See also page 44.) Evergreen **pyracanthas** are best trained along a wall in order to show off their plump

MORE TREES AND SHRUBS FOR BERRIES *Cornus alba* 'Siberian Pearls • *Gaultheria mucronata* 'Mulberry Wine' •

Callicarpa bodinieri var. *giraldii* 'Profusion'

Paeonia cambessedesii

Iris foetidissima

Symphoricarpos × *doorenbosii* 'Magic Berry'

Arum italicum subsp. *italicum* 'Marmoratum'

berries in scarlet, orange and yellow (see pages 125–26). Evergreen, acid-loving **gaultherias** (formerly pernettyas) provide some of the brightest berries in glowing pink and deep purple (see page 115).

The deciduous shrub ***Callicarpa bodinieri*** also has bright purple berries which, once the leaves have fallen, are revealed clinging to the straight branches. The snowberry, ***Symphoricarpos albus***, produces some of the purest white berries, clustered together on the ends of the stems like fistsful of marbles. The birds seem to leave these berries until last, so they persist well through winter. They are very useful in wreaths and indoor decorations, although the shrub itself is rather insignificant, with wiry stems on a thicket-forming bush. It is best planted in a wild part of the garden, where it will grow in dense shade beneath trees. The ***Symphoricarpos* × *doorenbosii*** hybrids are more compact with larger berries, rose-pink in **'Magic Berry'**, and white flushed with rose in **'Mother of Pearl'**.

There are some perennials too that have colourful berries. ***Iris foetidissima***🏆 is an undemanding evergreen that often gets relegated to dry shade, where relatively little else will flourish. It bears insignificant pale yellow and mauve flowers in summer, but these develop into fat seed pods that, in autumn, split open to reveal shining orange berries. These remain on the plant well into winter, although it is a great temptation to pick them for autumn flower arrangements. There is a variegated form with cream-striped leaves that show up well in winter shade but it seldom flowers, so no seed pods. In recent years a virus seems to be affecting these irises, causing the leaves to wither away in summer. The plants that survive produce seedlings that seem healthy but are struck down in adulthood. Some peonies, such as ***Paeonia cambessedesii***🏆 and ***Paeonia broteroi***, also have brilliant carmine seed cases, splitting open to release polished black seeds; and ***Arum italicum* subsp. *italicum* 'Marmoratum'**🏆 has succulent glistening orange berries before its leaves (see page 95).

BIRDS IN THE GARDEN

Birds are a very welcome addition to the winter garden and pleasure is to be had from watching their antics through the windows. They need to be encouraged by providing plenty of food, shelter and hiding places, especially if there are prowling cats.

Planting a good range of trees and shrubs will give birds a variety of habitats to use as nesting places, song perches and safe areas for watching for predators. Providing a wide selection of food sources is equally important, and our gardens, with a good mixture of plants, are becoming increasingly valuable in this respect, as farmland has turned over to monoculture. It is also essential to encourage the insects and grubs on which many birds feed, and these will need nectar-rich plants such as eupatoriums, sedums and buddleias.

Berry-bearing shrubs and trees are an asset in the garden but you must be prepared to share some of the fruit with the birds. If you want holly berries for Christmas decorations, then it is wise to cut some stems and store them in a cool place before the birds start to strip the branches. Trees such as sorbus and amelanchier provide plenty of attractive berries in autumn, and in spring their blossom will attract aphids and other insects on which nesting birds can feed; honeysuckle and other flowering climbers are also a good source of food.

Seed-eating birds are happy to feast on the seed of herbaceous plants that have been allowed to stand through the winter. *Phlomis russeliana*, with its stiff stems bearing whorls of seedheads all the way up, is an excellent plant not only for the winter gardener but also for the birds. Many grasses also produce abundant seeds, and small birds are adept at extracting the last morsel from the swaying stems.

It is often surprising to discover how many birds have been nesting in a garden, particularly in deciduous shrubs, as the nests only become apparent when the leaves have fallen, having been cunningly hidden in the summer. Conifers and evergreen shrubs are important, too, providing shelter and protection in the winter. Ivy is wonderful for birds, not only giving good cover but producing abundant black berries in the autumn on mature plants.

Putting out birdseed, nuts and fat balls will further encourage birds to visit, and there are many and various feeders from which to choose. If you do start feeding, be sure to do it regularly, as the birds will come to rely on the food you provide. They also appreciate a supply of water that can be kept free of ice, not only for drinking but for bathing as well.

A garden full of birds will need fewer chemicals to keep pests at bay, as small birds eat a prodigious number of aphids and other troublesome creatures, some birds deal with snails, and even ants are eaten. They will also fill the garden with birdsong and bring movement and colour to the winter scene.

PLANTS FOR ORNAMENTAL FRUIT

Amelanchier
Arbutus
Cornus (dogwood)
Cotoneaster (**1**)
Crataegus (hawthorn)
Euonymus europaeus
Hedera (ivy)
Ilex (holly)
Leycesteria formosa
Lonicera (honeysuckle)
Mahonia
Malus (crab apple)
Pyracantha
Rosa moyesii
Ruscus aculeatus (butcher's broom)
Sambucus (elder)
Skimmia
Sorbus
Symphoricarpos (snowberry)
Vaccinium
Viburnum opulus (guelder rose)

PERENNIALS FOR SEEDHEADS

Achillea
Aster
Centaurea
Crocosmia
Dipsacus fullonum (teasel)
Echinacea purpurea
Echinops ritro
Eryngium
Lythrum salicaria (purple loosestrife)
Phlomis (**2**)
Sedum spectabile
Verbascum

SEEDHEADS

The faded flowerheads and stems of many perennials provide warm tones in the winter border and are spectacular when etched by frost.

The seedheads on many plants are a great feature and those that are sufficiently robust to survive winter winds often linger for months. Delay cutting back some of those herbaceous plants that will last for a while in early winter. The small-flowered species asters, *Aster lateriflorus* and *Aster ericoides*, for example, are enchanting when a hoar frost catches their faded daisy discs. The feathery plumes of spent astilbe flowers turn russet, and the bold mahogany-brown seedheads on some of the rodgersias are very striking; *Rodgersia pinnata* 'Buckland Beauty' and 'Maurice Mason' hold their colour particularly well. The flowers on the sedums, too, fade to a warm foxy brown and remain on the stalks when the leaves have withered, their flat shape showing up well (see pages 22–23). The prickly seedheads of the thistle-like *Eryngium variifolium* are offset against the winter rosettes of green leaves veined with white.

Phlomis russeliana🏆 is a most useful plant with large, heart-shaped leaves, soft to the touch, which can cope with quite dry soil and sun or shade; in time it will grow into a weed-smothering mat. In summer the stiff flowering stems emerge, growing to about 1m (3ft) high, clothed with whorls of hooded, dark yellow flowers. Gradually the flowers fade and the seedheads develop, turning biscuit brown and remaining on the plant. They are wonderfully attractive rising above the overwintering basal leaves and look spectacular topped by a fall of snow, like icing on a bun (see page 28). They are so sturdy that the worst of the winter weather does not damage them, and it is hard to decide when finally to cut them down as they still act as a foil for spring-flowering plants like Bowles' golden grass, *Milium effusum* 'Aureum'🏆, and blue- and white-flowered brunneras. There is another herbaceous phlomis, ***Phlomis tuberosa* 'Amazone'**, with narrower leaves and lilac flowers carried on taller stems, which makes a splendid vertical accent in a dry sunny border, rising above carpeting plants. The seedheads are smaller, more like bobbles all the way up the stem, but also eye-catching when frosted.

Phlomis tuberosa 'Amazone'

Lunaria annua

The dried seedheads of honesty, ***Lunaria annua***, dangle like paper pennies from slender stems, adding a silvery sparkle to winter planting schemes. The flowers, borne in late spring and early summer, are usually purple, but there is a white form, ***Lunaria annua* var. *albiflora***🏆, that is especially beautiful. It is best planted at the back of borders among deciduous shrubs, where the white flowers will shine in spring. Honesty does not mind dry shade, and it is easy to remove any unwanted seedlings that spring up in the wrong place.

Calamagrostis brachytricha

Stipa tenuissima

Anemanthele lessoniana

GRASSES AND SEDGES

It is only in recent years that grasses and sedges, in a variety of leaf colours, have been recognized as excellent winter-interest plants and are now being used to great effect in our gardens. Many deciduous grasses fade in the autumn to delicate shades of caramel, biscuit and straw; some evergreen ones, really everbrowns, deepen in colour to a rich mahogany, while others come in shades of blue, grey and olive green. Planted in a prominent spot where they can be seen from the house, grasses and sedges are sure to catch the attention, offering texture, form and even movement.

Many grasses that flower in late summer and autumn hold on to their flowers into the early part of winter, fading to shades of straw and coffee. Some of the **calamagrostis** are particularly good at withstanding the ravages of winter weather and they last so well that is it hard to find the courage to cut them down, which is necessary before the new growth starts in the spring.

Calamagrostis* × *acutiflora* 'Karl Foerster'** makes a wonderful vertical silhouette, with the flowering heads held bolt upright on thin stems rising to 1.5m (5ft), and remaining upright when many other plants have fallen over. ***Calamagrostis brachytricha is more graceful and arching, with substantial bottlebrush-like flowerheads ripening in tones of green and mauve, changing with the onset of frosty weather to a mix of orange and cinnamon, and finally softening to fawn. The tenacious foliage copes well with winter gales; in summer the arching mound of narrow green leaves, up to 1m (3ft) high, contrasts well with low-growing perennials.

The very similar **stipas** are good winter plants, especially the elegant little ***Stipa tenuissima***, which wafts in the slightest breeze. Individual plants are not long-lived but they happily seed around to leave replacements. A frosted puff of *tenuissima* is a winter treasure. Altogether taller, rising to a majestic 2m (6ft), is ***Stipa gigantea***♡, with its fan of creamy-buff stems forming a see-through curtain, but this grass does get damaged by wind. ***Anemanthele lessoniana*** (formerly *Stipa arundinacea*) is a great find for light shade, with narrow arching leaves about 60cm (2ft) high, developing tones of warm orange, tan and even red in the winter months.

OTHER GRASSES AND SEDGES FOR WINTER *Acorus gramineus* 'Ogon' • *Arundo donax* var. *versicolor* •

Deschampsia cespitosa is another good grass, with dense tussocks, 60cm (2ft) high, of narrow evergreen leaves, rich green in summer, fading to straw. From these arise the flowering stems topped by loose, airy panicles, green at first, turning bronze and gold in several named cultivars such as **'Bronzeschleier'** (BRONZE VEIL) (see page 121) and the smaller **'Goldtau'** (GOLDEN DEW). Although deschampsias are happy in sun, they also cope with shade and make excellent partners for hellebores and ferns. They can be used in mass planting under trees, giving an airy effect among clumps of robust perennials such as persicarias and echinaceas. Another, smaller grass for shade is ***Chasmanthium latifolium***, with broad-bladed leaves and flattened oat-like flower spikes dangling from the nodding stems.

There is an enormous range of the fine ornamental grass **miscanthus**. One of the best for winter architecture is ***Miscanthus sinensis* 'Variegatus'**♀. It does not flower freely, but it forms a slender, bamboo-like column, 1.5m (5ft) tall, with curling leaves that are boldly striped green and white in summer, but by early winter have bleached to pale straw. ***Miscanthus sinensis* 'Gracillimus'** has finer foliage and a lighter habit – a graceful fountain flowing over low-growing plants.

The Japanese mountain grass, ***Hakonechloa macra***, arches out from a central mound about 60cm (2ft) high. The leaves may be green, gold or variegated, turning biscuit brown in autumn and remaining through the winter (see page 101). It is superb flowing over the edge of a pot, and mingles well with other plants.

The many cultivars of ***Molinia caerulea*** (purple moor grass) are arresting in autumn, their good-sized clumps of variously coloured leaves and dainty seedheads fading in a blaze of glory from honey gold to pale buff, but they do get damaged by winter gales.

The evergreen New Zealand sedges are perfect year-round plants, which can be

Miscanthus sinensis 'Variegatus'

used to grace many areas of the garden. There are several brown varieties, one of the best being ***Carex comans* 'Bronze'**, a low tussock of pinkish bronze in the summer, deepening a tone in winter. It can be planted to grow through a carpet of the evergreen, cream-variegated perennial *Arabis alpina* subsp. *caucasica*, or next to the trunk of the mahogany-barked cherry, *Prunus serrula*♀ (see page 144). ***Carex buchananii***♀ is similar, with narrower leaves and a more vertical habit (see page 101), while ***Uncinia rubra*** makes a low mound of much stiffer leaves, of glowing ruby red in good forms.

Clumps of ***Carex dipsacea*** are up to 45cm (18in) high, with leaves in shades of olive green, tan and ochre; these make an excellent companion for some of the new heucheras with orange and caramel leaves, such as *Heuchera* 'Amber Waves', CRÈME BRÛLÉ ('Tnheu041') and 'Marmalade' (see page 59). These sedges take on the strongest colour in sun and, combined with a sharp acid-green shrub such as

Molinia caerulea 'Variegata'

Carex comans 'Bronze'

Uncinia rubra

Hebe rakaiensis♀ (see page 89), bring an unusual colour to the winter scene.

Some sedges have variegated foliage, such as ***Carex conica* 'Snowline'**, with a white stripe down each narrow green leaf (see page 95), and ***Carex oshimensis* 'Evergold'**♀, with slightly broader leaves striped in gold. Growing in clumps up to 30cm (12in) high, they are excellent planted next to hellebores in maroon and cream. ***Carex comans* 'Frosted Curls'** is a similar size, with twisting, thread-like leaves in pale green and cream.

Carex morrowii 'Variegata' • *Cortaderia selloana* 'Pumila' • *Luzula sylvatica* 'Aurea' • *Miscanthus sinensis* 'Flamingo' •

Approaching spring

The weather is warming up, the days are getting longer and the light better, and more and more plants are bursting into growth. As winter moves into spring, promising buds begin to appear, the stems on many twiggy shrubs are tipped with colour, and the new shoots of herbaceous plants are starting to push up through the soil.

Magnolia stellata

Salix caprea

One of the first signs that spring is on its way is the buds beginning to break on trees and shrubs. The soft furry buds on **magnolias** just ask to be stroked, and the silky buds on the pussy willows (***Salix caprea***) open into fluffy catkins. The flowering currants, ***Ribes sanguineum***, carry pink and red buds set off by the newly unfurling pleated green leaves. On many twiggy shrubs the shoots are tipped with fresh green, lime yellow and even orange on some ***Spiraea japonica*** forms. Graceful ***Amelanchier lamarckii***🏆 has tiny, suede-like emerging leaves in coppery-pink, soon to be followed by showers of starry white flowers.

Herbaceous plants, too, are beginning to emerge and here some of the shoots are dramatic indeed, not only the bright green of daylilies but the brilliant red of peonies and the scarlet of the ornamental rhubarb ***Rheum* 'Ace of Hearts'**. As early as late winter the red buds of ***Paeonia mlokosewitschii***🏆 thrust through the soil, slowly elongating into asparagus-shaped spears in a rich rhubarb red, forming an exotic contrast with the last of the snowdrops and hellebores. Many euphorbias come with colourful new shoots: ***Euphorbia griffithii***

Ribes sanguineum

Spiraea japonica 'Goldflame'

in orange, and ***Euphorbia schillingii*** in maroon, with sea-green leaves edged with red and centred with cream.

While the common primrose, ***Primula vulgaris***, may produce the occasional flower from midwinter, only in early spring does it comes into full production, flaunting its soft yellow flowers surrounded by crinkled leaves wherever it has been able to seed itself. It often moves around the garden, seldom remaining in the same spot for more than a couple of years, preferring to seed into fresh soil.

Crossing over from winter into spring are the fleeting flowers of the little perennial ***Cardamine quinquefolia***, with a flush of fresh green palmate leaves with jagged edges and trusses of pinkish-mauve, four-petalled flowers. It happily hides the dying foliage of snowdrops and, as it prefers to be in shade, can also be grown around the emerging leaves of hostas and the unfurling fronds of ferns. Some of the small narcissus are now starting to show their flowers, including ***Narcissus* 'February Gold'** and ***Narcissus* 'Tête-à-tête'** (see page 66). But beware – winter can suddenly reappear and some plants may have their top growth seared off.

***Veronica peduncularis* 'Georgia Blue'** is the perfect plant to take us from winter into spring, forming a useful carpeting creeper with small ferny leaves tinted bronze in winter and turning green by summer. The periwinkle-blue flowers each have a smart white eye and they smother the plant in spring. 'Georgia Blue' likes full sun and good drainage, and if happy will wander around the base of shrubby plants such as the glaucous-leaved euphorbias or hebes, the blue flowers peeping out from beneath their stems. Given space it will climb over rocks.

Brunneras are now flowering earlier each year, almost in winter, and they seem set for a great future as flowering foliage plants. Their delightful forget-me-not blue flowers appear first, to be followed by large heart-shaped leaves; in some

Paeonia mlokosewitschii

Cardamine quinquefolia

Euphorbia schillingii

Veronica peduncularis 'Georgia Blue'

Primula vulgaris

Stachyurus praecox

varieties these are plain green, in others they are sprinkled with aluminium spots or are variegated in cream and several shades of green; in some newer varieties such as ***Brunnera macrophylla* 'Jack Frost'** and **'Looking Glass'** they are almost completely silver. They grow happily in shade, and their rough, hairy leaf surface is unpalatable to slugs and snails, so the colourful foliage remains in pristine condition throughout the summer.

***Skimmia japonica* 'Rubella'**, which has been brightening containers with its claret buds all winter, now opens its clusters of creamy-white flowers with a spicy scent; other skimmias are also coming into flower (see page 49).

The beautiful but rarely planted shrub ***Stachyurus praecox*** finally flowers: all winter the stiff, bare branches have carried mahogany-brown buds in downward-pointing spikes, and these now open into pale yellow, bell-shaped flowers, hanging like tear drops all along the stems. It is best planted against a wall where the flowers will show up well.

Planning for all seasons

Space is precious in any garden and it makes little sense to grow too many winter-flowering plants that are not worth a glance in the summer. Far better to plan for interest throughout the seasons, using a mix of small trees and shrubs – including evergreens of various colours – along with plenty of perennials, bulbs, annuals and biennials. There are some plants that do look good all year, perhaps providing attractive foliage, fine flowers and handsome berries, and these are worth their weight in gold.

Above: Evergreens, including *Euonymus fortunei* 'Emerald 'n' Gold' and grasses, maintain the interest in this border in winter. Below: The white trunk of *Betula utilis* var. *jacquemontii* provides a focal point.

ALL-YEAR PLANTINGS FOR WINTER SHRUBS

There are a few plants that earn their space in the garden in all four seasons and of these the deciduous dogwoods (***Cornus***) are some of the best performers. Their prime attraction is their brightly coloured stems in winter, coming in reds, oranges and yellows and even dark purple. In spring the newly unfurling leaves are pale lime green, overshot with bronze, and in some the green summer leaves are margined with cream or yellow; there are also cultivars with pale gold leaves. In autumn when the frost comes these leaves flare up in red and purple before they fall, revealing the coloured stems. Few plants indeed are capable of such a sustained effect. Two of the best to consider are ***Cornus sanguinea* 'Midwinter Fire'** (see page 148) and ***Cornus alba* 'Sibirica Variegata'**♀, which has leaves with a broad creamy-white margin, colouring

well in the autumn. In both cases half the stems should be hard pruned each year in late spring, to encourage new stems that will look good the following winter. Their open branchwork in winter shows best against a strong background, so an evergreen shrub that will give structure and interest throughout the year is a natural choice; possible partners include *Brachyglottis* 'Sunshine'♡, *Osmanthus* × *burkwoodii*♡ and *Choisya ternata*♡ (see pages 88, 89 and 105).

Cornus mas **'Variegata'**♡ is a less vigorous form of the cornelian cherry, making a small bushy tree or rounded shrub with yellow pompom flowers festooning the bare branches in late winter. It has pointed leaves with a wide white margin and in autumn bears scarlet fruit like small cherries. This would combine well with an underplanting of yellow aconites in winter to echo the colour of the flowers, followed by blue wood anemones, particularly the large-flowered *Anemone nemorosa* 'Buckland', and then the small greenish-gold *Hosta* 'Hydon Sunset'; the gently spreading *Galium aristatum*, running through all of these, would give starry, scented white flowers in summer.

Viburnum **×** ***globosum*** **'Jermyns Globe'** is a valuable evergreen shrub, providing interest all through the year. In time it forms a dense, rounded, medium-sized bush with narrow, pointed, leathery dark green leaves on reddish stalks. It makes the perfect background for the white-stemmed Himalayan birch *Betula utilis* var. *jacquemontii*♡ (see page 145), the viburnum's dark green leaves intensifying the whiteness of the birch bark in all seasons. In late spring the viburnum is decorated with clusters of small white flowers opening from pink buds. The shrub's compact, rounded habit makes it ideal as a host for an autumn-flowering viticella clematis; *Clematis* 'Alba Luxurians'♡ is a good choice, the green-tipped white flowers showing up well against the green leaves; it is a vigorous plant, but the viburnum is strong enough to support it. In early winter the clematis can be cut back to ground level so that its dead leaves do not disfigure its host. Winter aconites complete the picture, growing at the base of the birch.

Above: In summer, flowering perennials and richly varied foliage add colour to the border pictured on the opposite page. Right: The dark leaves and white flowers of *Viburnum* × *globosum* 'Jermyns Globe' are the ideal background for the white bark of the birch shown on the opposite page.

If thinking ahead to summer flowers, also consider using winter evergreens to complement them. Gold and yellow-variegated shrubs make a happy combination with yellow flowers in spring and summer. The popular ***Euonymus fortunei*** **'Emerald 'n' Gold'**♡ (see page 98) blends with yellow-flowering hypericums, the shrubby *Potentilla fruticosa* 'Elizabeth' and perennials such as *Anthemis tinctoria* 'Sauce Hollandaise', yellow lupins and the tall *Helianthus* 'Lemon Queen'♡. The light airy evergreen ***Lonicera nitida*** **'Baggesen's Gold'**♡ (see page 96), with tiny, golden yellow leaves that turn a lighter yellow-green in cold weather, is a star in the garden in winter.

The gold-variegated foliage of *Euonymus fortunei* 'Emerald 'n' Gold' mixes well with white lychnis and a yellow potentilla in summer; it also continues to provide colour in winter (see the picture opposite).

In a border full of pastel perennials, a white- or silver-variegated evergreen shrub would provide a background in summer and colour and structure in winter.

It is also a good mixer with summer flowers, making a fine companion for the pure yellow *Rosa* MOLINEUX ('Ausmol')🏆 and the sapphire blooms and red autumn foliage of *Ceratostigma willmottianum*🏆.

In the silver and cream spectrum a variegated holly such as ***Ilex aquifolium* 'Handsworth New Silver'** (see page 93) is a perfect backdrop for white border phlox, blue and white *Campanula persicifolia,* blue-grey *Aconitum* 'Stainless Steel' and blue hardy geraniums like *Geranium pratense* 'Mrs Kendall Clark'🏆 or the twice-flowering *Geranium pyrenaicum* 'Isparta'. The silver-variegated ***Pittosporum* 'Garnettii'**🏆 (see page 92) makes an excellent structure shrub alongside the ever-popular white *Rosa* ICEBERG ('Korbin')🏆 and *Buddleja* 'Lochinch'🏆, with silver foliage and blue flowers in late summer. The foreground could be planted with blue forms of the pretty creeping violet *Viola cornuta*🏆, *Lavandula* × *chaytorae* 'Sawyers'🏆 and perhaps clumps of blue muscari and white *Narcissus* 'Thalia'🏆 for spring colour.

PLANTINGS FOR WINTER PERENNIALS THROUGH THE SEASONS

Hellebores (see pages 72–73) are ideal plants to place at the back of a border, tucked among deciduous shrubs. Here the flowers can be enjoyed in winter and their mounds of dark green leaves will cover the ground beneath the shrubs and keep unwanted weeds at bay. They are happy beside flowering shrubs such as deutzias and weigelas or complementing foliage plants like *Physocarpus opulifolius* 'Dart's Gold'🏆 (see page 148). For a more prominent position consider the green-flowered ***Helleborus multifidus* subsp. *hercegovinus***, which has finely divided leaves, almost fern-like in appearance; it makes a delightful summer companion for blue and white forms of *Viola cornuta*🏆.

The perennials that retain their leaves in winter often go on to flower in the spring or summer and provide foliage contrast

WHERE TO GROW WINTER PLANTS

A decision has to be made when thinking about winter planting. For maximum impact you can dedicate a corner of the garden to winter plants: a small area with *Viburnum farreri*🏆 planted with hellebores and snowdrops at its feet and Christmas box (*Sarcococca*) nearby for scent would suffice; the drawback here is that this area would not warrant a second glance for the rest of the year. The alternative is to scatter the winter charms around the garden: they will need to be sought out, but will then offer pleasant surprises on a winter walk.

WINTER IN SUMMER

A bed full of winter-flowering hellebores (*Helleborus* × *hybridus*) is a delight from midwinter (**1**) right through to the middle of spring. The handsome foliage remains to provide interest through the rest of the year, but it is a shame not to have something to give colour in the summer months. One suggestion would be to grow daylilies (*Hemerocallis*) in among the hellebores: they start to bloom in early summer (**2**), with flowers in shades of yellow, orange, pink, and cherry red opening in succession for many weeks, and their sword-like, bright lime-green new foliage would make a fine contrast with the darker green, palmate leaves of the hellebores early in the spring.

in summer borders. Many **bergenias** are grown for their richly coloured winter leaves (see pages 54–55), but they also produce sugar-pink to ruby-red flowers in spring; by this time new leaves are coming through, in a pleasing mid-green, and these in turn are a foil for plants such as the cone flower, *Echinacea purpurea*, with stiff branching stems bearing pinkish-crimson daisy flowers in autumn.

Epimediums are excellent plants for dry shade but need not be banished to some out-of-the-way corner; they make attractive mounds of foliage, green edged with copper in the new leaves, turning to bright green later (see page 55). In summer, they look pleasing combined with oxonianum geraniums, such as *Geranium* × *oxonianum* 'Walter's Gift', which has pale salmon flowers and distinctive foliage heavily marked with chocolate. Add *Astrantia* 'Hadspen Blood', a charming plant with dainty, dark red flowers, and a white-flowered *Viola cornuta*♛. Pots of the lovely *Lilium regale*♛ dropped in behind will provide elegant white trumpet flowers with a heady fragrance.

Some of the **heucheras** with richly coloured foliage also flower in summer, and plants with purple leaves and pale pink flowers or orange leaves and cream flowers merit a place in the summer border (see pages 58–59). ***Heuchera* 'Marmalade'**, for example, makes a vibrant planting partner for the purple-leaved *Ligularia dentata* 'Britt-Marie Crawford', which has deep orange flowers, and ***Heuchera* 'Purple Petticoats'**♛ associates easily with *Geranium* 'Mavis Simpson'♛, with soft pink flowers all summer long. A purple heuchera would also look well in front of the dwarf hardy *Fuchsia* 'Tom Thumb'♛, with dangling crimson and purple flowers, or next to the blue-leaved *Dianthus* 'Mrs Sinkins', which has double white flowers with fringed petals.

The invaluable ***Tellima grandiflora* Rubra Group**, with leaves that turn coral red in winter (see page 60), has pleasing greenish-cream, little bell-like flowers in early summer. It combines happily with the silver-edged foliage of *Pulmonaria* 'Mary Mottram' or the chubby pink spires of *Stachys officinalis* 'Rosea Superba'.

The amber leaves of *Heuchera* 'Marmalade', together with hosta, brunnera and spiraea, partner the wine-black foliage of *Ligularia dentata* 'Britt-Marie Crawford'. The heuchera provides colour in winter after the other perennials have faded. Hellebores and early bulbs could fill any gaps in the border.

FILLING EMPTY SPACES

Many summer-flowering herbaceous perennials grow from a neat central core, often covering quite a lot of bare earth with trailing flowering stems, which then die back in winter. Rather than waste this space, use it for small winter-flowering bulbs, many of which will welcome being lightly shaded by the perennial's foliage in summer.

For a combination that looks good virtually all year, plant a few snowdrops and perhaps *Cyclamen coum*♛ around a hardy geranium such as *Geranium* × *oxonianum* 'Spring Fling', which has attractive green, cream and brown leaves in the spring followed by masses of small pink flowers throughout summer (above).

WINTER-FLOWERING SHRUBS IN SUMMER SHADE

Many winter-flowering shrubs – for example, *Viburnum farreri*♛, winter honeysuckles such as *Lonicera* × *purpusii* and *Rhododendron dauricum* (right) – while lovely when in bloom, do not merit a second glance after their flowers have faded. Fortunately they are happy growing in shade, where shrubs that bloom in summer simply would not thrive. Partnering them with evergreen shrubs with prettily variegated leaves maintains interest all through the seasons.

ALL-SEASONS FLOWER BEDS

NEAR THE HOUSE

The following is a planting scheme for a shady bed, backed by a high wall, in the author's garden. It is sited close to the house, and the changing display of flowers and foliage can be seen through the windows month by month. The pictures show the bed as it appears in late winter (**1**), early spring (**2**) and summer (**3**).

1

2

3

Against the wall grow × *Fatshedera lizei*🏆, for its dramatic evergreen glossy leaves, and the Christmas box *Sarcococca hookeriana* var. *digyna*🏆, with its small, tassel-like, pinkish-white flowers, for unexpected scent in late winter. In the bed itself the season starts with a very early-flowering selection of the Christmas rose, *Helleborus niger*🏆, and a dark purple *Helleborus* × *hybridus* tucked at the base of the fatshedera, along with several plants of the winter-flowering *Cyclamen coum*🏆, with silver-marbled leaves and pink and white flowers. Snowdrops, starting to appear in late winter, include some more unusual forms, such as *Galanthus nivalis* 'Tiny', which is only 10cm (4in) high, as well as the later-flowering *Galanthus ikariae*, with broad-bladed, bright green leaves and quite large flowers on short stalks. To follow there is a dramatic version of the winter aconite, *Eranthis hyemalis* 'Guinea Gold'🏆, with large bright yellow flowers rising from a ruff of bronzy-green foliage; this nestles alongside the dark chocolate-brown leaves of the celandine *Ranunculus ficaria* 'Brazen Hussy', now starting to appear, and the tiny yellow and green flowers of *Hacquetia epipactis*🏆.

With the winter plants dying down, in the spring *Corydalis solida* claims attention, with its grey-green ferny leaves and pretty mauve flowers, as does *Mertensia virginica*🏆, a woodland perennial with clusters of trumpet-like flowers in bright blue. These plants themselves die down in early summer. *Primula* 'Guinevere'🏆 provides a splash of pink, and the unfurling fronds of the Japanese painted fern, *Athyrium niponicum*, add soft silver-grey, with a touch of red in the veins. For early summer, there is the arum-like *Arisaema candidissimum*, whose stunning spathes, striped in pink and white, are followed by dramatic leaves, each divided into three rather like a giant trillium. These arch beautifully over a small blue-leaved hosta, the summer-flowering *Dicentra* 'King of Hearts', with its delicate glaucous leaves and dangling lockets of rich crimson, and a silver-leaved pulmonaria. All are threaded through with the cerise-red, black-eyed flowers of *Geranium* 'Sue Crûg' on wandering stems.

Continuing into the autumn are *Persicaria milletii*, with small crimson, bottlebrush flowers, and *Astilbe* 'Inshriach Pink', with feathery plumes of pale pink flowers and ferny leaves flushed with purple. Then, after a pause in late autumn, giving time for a tidy up and mulch of garden compost, the buds of the Christmas rose will begin to develop, starting the cycle again.

FRONT GARDEN

The following scheme is for a small sunny bed in the author's front garden. It is just outside the sitting-room window, so the planting is kept low. The pictures show the bed as it appears in winter (**1**), early spring (**2**) and early summer (**3**).

In winter the season starts with the Christmas rose (*Helleborus niger*🏆), with beautiful cream cups fading to pink. A neatly clipped bun of silver-variegated box, *Buxus sempervirens* 'Argenteovariegata', adds a touch of formality. Snowdrops are scattered all through the bed, some pushing up through the crimson leaves of *Tellima grandiflora* Rubra Group, and pink and white *Cyclamen coum*🏆 flower among the black grassy leaves of *Ophiopogon planiscapus* 'Nigrescens'🏆. *Bergenia* 'Baby Doll', with large paddle-shaped leaves, adds contrast, as do the tussocks of the coppery-pink sedge *Carex comans* 'Bronze'. In spring, the early-flowering *Anemone blanda* 'White Splendour'🏆, with pink-tinted ivory flowers, pops up in several places, and then the delicate *Dicentra* 'Stuart Boothman'🏆 appears, with pewter-grey ferny leaves. This bears numerous pretty pink locket-shaped flowers in early summer and then dies down, leaving space for a dwarf hardy fuchsia, *Fuchsia* 'Tom Thumb'🏆, no more than 50cm (20in) high, with an endless succession of carmine and purple flowers.

A pink hybrid musk rose, *Rosa* 'Felicia'🏆, takes centre stage in high summer, followed in the autumn by *Caryopteris* × *clandonensis* 'Worcester Gold'🏆, with rich blue flowers and pale gold foliage on a compact, low-growing bush. With the onset of colder weather, the splendid grass *Calamagrostis brachytricha* turns shades of orange and cinnamon and *Stipa tenuissima* wafts pale fawn fronds in the wind.

The bed is bounded by a picket fence, along which is trained *Chaenomeles speciosa* 'Moerloosei'🏆, with apple-blossom flowers in spring. This in turn hosts a herbaceous clematis, the lavender-blue *Clematis* 'Arabella'🏆, which gives a continuous display all through summer.

1

2

3

Climate change

Whether or not it can be ascribed to global warming, our climate does currently appear to be undergoing a period of change, with warmer, often wetter winters and hotter, drier summers. Perhaps most disturbing is that the climate seems so much more unpredictable, swinging from one extreme to the other in only a very short time. While changing conditions may mean gardeners will be able to grow a greater range of plants than ever before, we may also need to think again about how we grow them.

Scenes of the garden hidden by snow may become increasingly rare, given the changing climate.

There seems little doubt that our winters have been warming up in recent years. Consistent records show increasingly early appearances of snowdrops and narcissus, and winter aconites have been reported in flower before Christmas. The first leaves too are appearing early, with horse chestnut recorded in midwinter. This change in our weather may be only short-lived: some experts tell us we are heading for another ice age. But, for the time being at least, gardeners can enjoy the opportunity to grow a range of plants not previously considered fully hardy in Britain. Dahlia tubers can perhaps be left in the ground, winter-tender salvias may survive all year, and Australian and New Zealand shrubs such as callistemon, grevillea and leptospermum, given the right soil, may thrive outdoors.

Milder autumns and a tendency towards fewer cold nights, which would normally stimulate leaf fall, result in many deciduous and semi-evergreen shrubs hanging on to most of their leaves into midwinter. Roses and semi-evergreen honeysuckles are good examples. This does mean there is more green foliage in the winter garden, but in winter-flowering shrubs, lingering foliage can detract from the display. *Viburnum × bodnantense* and *Lonicera × purpusii* are two examples of winter bloomers that hang on to shrivelling leaves, which can conceal their flowers. If frost kills leaves before they fall, they can remain on the plants throughout winter. This can happen with witch hazels, the dead foliage spoiling the display of midwinter flowers; on small specimens, the leaves could be removed before the flowers open.

It is important to remember that there will still be the risk of hard frosts and these can be more harmful when plant growth has not really shut down for the winter, encouraged to keep going by a warm autumn. In an early spring, when plants struggle into growth prematurely they may be hit by a late frost. The lovely perennial *Dicentra spectabilis*♈︎, known as Dutchman's breeches or bleeding heart, is frequently a victim of this.

There are things we can do in the garden to help. We often lose plants not because of the severity of the weather but because we are trying to grow them in the wrong conditions. In their natural habitat, many alpine plants are normally frozen all winter beneath snow, and thus kept dry. In wetter climates, alpines will really benefit from extra drainage in the winter: grow them in slightly raised beds with an edging of flat stones, and fill in with gritty soil mixed with a little humus, such as leaf mould or garden compost, topped with a layer of gravel. Many early spring bulbs and tubers need to remain dry in summer when they are resting or they will rot. The solution is to plant these bulbs beneath deciduous trees, where the tree roots will help to keep them dry and there will be enough light and water for them to flower early in the spring.

Fortunately there are some splendid plants that are well able to cope with changing conditions. The hellebores are happy to come into growth and flower early in the year in mild weather; if the weather does change, with a long cold spell or even snow, they withdraw the sap from their stems and lie flattened on the ground until the snow melts or the sun returns. The marbled-leaved *Arum italicum* subsp. *italicum* 'Marmoratum'♈︎ has the same capability. Many early bulbs have developed a technique of suspended animation: snowdrops will start to appear in a mild spell, only to stop all growth in a cold snap, and then continue with their growth points undamaged, and they seem able to do this several times. Many of the early narcissus also behave in this way.

Mild autumns can encourage summer shrubs to continue flowering. The musk rose *Rosa* 'Felicia' may produce occasional blooms even in midwinter.

Some deciduous shrubs, such as witch hazel (*Hamamelis*), may retain their dying leaves, which can spoil the show of exquisite winter flowers.

Modern-day plant hunters are seeking hardier forms of some of the more tender plants that we are now trying to grow, searching out varieties that grow at higher altitudes and are thus better adapted to coping with periods of extreme temperature at both ends of the scale. More eucalyptus species from Australia are surviving outside in Britain, and perennial salvias from high in the Mexican mountains are being introduced. The process of natural selection is also already at work, with some plants of the borderline hardy perennial *Verbena bonariensis*♈︎ coming through the winter, setting seed and passing on their improved hardiness to their offspring.

We may have to continue mowing our lawns in late autumn, and adapt our methods of growing certain plants but, to compensate, we can take pleasure in the roses lingering on into early winter, greet the first spring bulbs ever earlier in the year, and enjoy an ever-widening choice of plants for the winter garden.

BERGENIA DIE-BACK

Some plants actually need a cold period early in the winter to start building up their frost protection. Bergenias are usually very tough plants, surviving extremely cold spells, but they need to get their natural anti-freeze working at the onset of winter. This inbuilt protection manifests itself as a red coloration of the foliage: pigments develop in the leaf to safeguard the normal functions of the plant. In mild autumns the leaves fail to colour, so when frost does arrive it damages the plants, causing the foliage to die back and the rhizomes to rot.

EVERGREENS

In winter plants with evergreen foliage in various colours and shapes are of prime importance. The many classic evergreen shrubs help to form the basic structure of beds and borders, with the addition of conifers and spiky architectural plants. In the foreground there are plenty of perennials with overwintering leaves, some green, others blue and grey, and many in shades of warm brown and glowing purple. These all provide the backdrop for the treasured flowers of winter.

RIGHT: *Helleborus argutifolius* leaves edged with frost.

Classic evergreens

Evergreen shrubs are a vital ingredient of the garden, helping to form the structural background, providing solid substance in winter and keeping the balance among the lighter deciduous foliage in summer. Plants with variegated leaves have become increasingly popular, particularly for lightening effects in the winter garden, and there are now many evergreens with cream, white, yellow and gold variations in the leaf colouring.

Ilex aquifolium 'Ferox'

The holly (***Ilex***) has been cherished in our gardens for centuries, bringing not only shining dark green, prickle-edged leaves but also bright red berries at a time when the autumn berries are going over. That is, if a male tree is nearby to pollinate the flowers on female plants so that they will bear fruit – and if you can manage to share the berries with the birds. Hollies can be grown as handsome specimen trees or trained in different topiary shapes, and holly hedges are very effective barriers, if somewhat painful to cut.

Of the many species and cultivars, ***Ilex aquifolium* 'Bacciflava'** is one of the

Ilex aquifolium 'Pyramidalis'

Ilex aquifolium 'Bacciflava'

Ilex crenata 'Convexa'

Ilex aquifolium 'Madame Briot'

Ilex crenata 'Golden Gem'

best for yellow berries, which are last to be eaten by the birds. The scarlet-berried ***Ilex aquifolium*** **'Madame Briot'**♀ has purple stems and spiny leaves edged with gold. These two form large conical shrubs, slowly reaching 5m (16ft) in height by 3m (10ft) across. ***Ilex aquifolium*** **'Pyramidalis'**♀, with bright green leaves and red berries, is also conical in shape, but somewhat narrower at only 1.5m (5ft) wide. It is self-fertile so does not need to be grown with a male for pollination. The hedgehog holly, ***Ilex aquifolium*** **'Ferox'**, is one of the oldest holly varieties, popular for its extremely prickly leaves, with spines not only on the edges but also growing from the upper surface. It is a small, neat, male bush that makes an excellent deterrent to people and animals (*ferox* means fierce). The Japanese holly, ***Ilex crenata***, is spineless and can be trimmed as a small bush or a low hedge. ***Ilex crenata*** **'Convexa'**♀ has black berries, as does ***Ilex crenata*** **'Golden Gem'**♀, which grows only about 60cm (2ft) tall, with small golden yellow leaves rather like those of box.

Hollies are not fussy about soil, although they will not be happy on heavy clay or in waterlogged ground, and they can cope with windy, exposed conditions and with atmospheric pollution. Neglected plants can be cut back into old wood to rejuvenate them, and the cultivars can be propagated from semi-ripe cuttings taken in late summer. With the variegated varieties, watch for plain green shoots and remove them as they appear (see page 93).

OTHER GOOD GARDEN HOLLIES

Although slow-growing in the early years, hollies make excellent structure plants for the garden. Some of the best include:

Ilex × altaclerensis **'Golden King'**♀ Almost spineless leaves with broad gold margins and masses of reddish-brown berries.

Ilex aquifolium **'Argentea Marginata'**♀ Green stems and broad dark green leaves with white margins. Plenty of scarlet berries. (See page 94.)

Ilex aquifolium **'Handsworth New Silver'**♀ Purple stems and long, grey-mottled dark green leaves with cream margins. Bright red berries.

Ilex aquifolium **'J.C. van Tol'**♀ Abundant, bright red berries and glossy, almost spineless leaves. (See page 26.)

Ilex aquifolium **'Myrtifolia Aurea Maculata'**♀ Dense and compact male variety; a good choice for the smaller garden. Dark green leaves with central splashes of gold.

Buxus sempervirens

The common box, ***Buxus sempervirens***♀, has been used for hedges and trained into many and various topiary shapes since medieval times. It is extremely easy to propagate from semi-ripe or semi-hardwood cuttings in summer and this is the best way to build up stock for hedging. There are plenty of cultivars to suggest: ***Buxus sempervirens*** **'Suffruticosa'**♀ is a dwarf, dense plant of a good rich green, normally only 30cm (12in) or so in height; ***Buxus sempervirens*** **'Elegantissima'**♀, which has silver-variegated leaves, is the perfect choice for a neat specimen bun shape, growing slowly to 1m (3ft); ***Buxus microphylla*** **'John Baldwin'** has small bluish-green leaves and a vertical habit that lends itself to shaping into cones and pillars; and ***Buxus microphylla*** **'Faulkner'**, with small emerald-green leaves, has a low, spreading habit that makes it ideal for low hedges and trimming into balls and broad cones.

Buxus sempervirens 'Suffruticosa'

Buxus sempervirens 'Elegantissima'

BOX BLIGHT

Box blight is becoming a worrying problem. Sometimes mistaken for nutrient deficiency, it appears as a browning of the foliage, spreading through the plant. The foliage eventually dies and turns parchment coloured. The disease is more prevalent in damp, humid conditions. Any affected growth should be cut out and the plants kept healthy with a soluble feed and plenty of water in dry weather. Plants usually recover although they are rendered unsightly for a year or two.

Variegated ivies make wonderful ground-cover plants for shade under trees and lighten the planting when combined with plain evergreens.

Buxus sinica is a very hardy species, particularly good in colder climates. The dark velvety green ***Buxus sinica* var. *insularis* 'Winter Gem'** will take years to grow into a mounded bush 1m (3ft) high by 1.2m (4ft) wide. ***Buxus sinica* var. *insularis* 'Justin Brouwers'** is even slower-growing, naturally forming a perfect little bun.

Common ivy, ***Hedera helix,*** is a self-clinging, evergreen climber, using trees and walls for support in growing upwards. It has two stages of growth. In the juvenile stage the climbing shoots develop aerial rootlets with adhesive suckers to help them to cling; when they reach the top of a tree or wall, they change to their adult form and cease to cling, and the leaves become more rounded and lose their lobes. At this stage the shoots will produce nectar-rich green flowers in autumn, followed by black berries. Semi-hardwood cuttings may be taken from the adult shoots and these will slowly grow into free-standing plants. Ivy is very adaptable, thriving in almost any soil and coping well with shade. It is not only a climber, but will also creep along the ground.

Hedera helix f. *poetarum* 'Poetica Arborea'

THE HOLLY AND THE IVY

One of the few British native evergreen trees, the holly has long been considered a magical plant. While other trees lost their leaves in winter, the holly remained bright and glossy, with shiny red berries, a symbol of hope during the long dark winter months. It was always considered very bad luck to cut down an entire tree, but branches were cut to decorate house and barn – and churches later on; tradition has it that they were never to be taken in before Christmas Eve otherwise a family quarrel would result. Hollies were planted near the house as protection against lightning and to keep witches away; they were left standing proud in a hedgerow, from the old belief that witches ran along the tops of hedges and a full-grown holly tree would stop them in their tracks. Sprigs of holly were hung up in barns to protect cattle from evil spirits, and the berries were made into a medicinal drink to cure coughs and colds.

The common ivy, *Hedera helix* (left), was another evergreen welcomed in darkest winter. Like holly, ivy was used in pagan rituals and in Europe it was associated with Bacchus, the Roman god of wine. In the Middle Ages ivy-covered poles, known as 'bushes', were used to indicate a tavern: the higher the pole the more important the establishment. Garlands of ivy decorated churches, and there are some fine medieval carvings of ivy in Westminster Abbey. Ivy leaves were sometimes fed to cattle as an emergency winter fodder, and the plant was also thought to have magical powers in protecting livestock.

There are many different leaf shapes, in plain green or in variegated patterns. ***Hedera helix*** **'Duckfoot'**♡ has tiny, pale green leaves shaped like a webbed foot and can be clipped into a neat mound 30cm (12in) high. ***Hedera helix*** **'Erecta'**♡ is architectural in form, with stiff, upright shoots and arrow-shaped dark green leaves. Very slow-growing, it will eventually reach 90cm (3ft) high. ***Hedera helix*** **f.** ***poetarum*** **'Poetica Arborea'** (poet's ivy), with yellow berries, is a selection of the adult ivy that will slowly develop into a rounded bush 2m (6ft) high. In winter, if grown in full sun, the bright green, shallowly lobed leaves turn burnished copper with prominent green veins.

The butcher's broom, ***Ruscus aculeatus***, is a small, tolerant evergreen shrub for dense shade. It is said that butchers would use it to sweep the floors of their shops. Today, it is most often seen as cut foliage from florists. What appear to be leaves are in fact tiny flattened stems, tipped with sharp points, in a dull, dark green. Unless you choose a hermaphrodite form, male and female plants will be needed for the cherry-like berries that follow the white spring flowers, and these berries last well over the winter. It is a very tough shrub, able to cope with the driest of shady sites, so is seldom given a better position where its berries would be seen more clearly. The related ***Danae racemosa***, with similar leaf-like stems (see Good Companions, below), is a hermaphrodite and sometimes produces berries after a hot summer.

Prunus lusitanica♡, the Portugal laurel, is an invaluable background evergreen shrub growing into a broad cone over 3m (10ft) high; it has neat, glossy, ovate leaves on red stalks. ***Prunus lusitanica*** **'Variegata'** has leaves in two shades of green, rimmed with white.

Ruscus aculeatus

Danae racemosa

Prunus lusitanica

Prunus lusitanica 'Variegata'

OTHER GOOD IVIES

Not all ivies are invasive; some compact forms are slow-growing and make excellent small specimens for planting in narrow borders or in pots and other containers.

Hedera helix **'Pedata'** Leaves divided into three lobes, the middle one being much longer than the others, like a bird's foot.

Hedera helix **'Green Ripple'** (above) Small, sharply lobed leaves, the central lobe being the longest. The rich emerald-green foliage looks good as a partner for snowdrops.

Hedera helix **'Spetchley'**♡ Tiny leaves, forming a dense, low mat. Excellent for planting in gravel.

GOOD COMPANIONS

The dark green foliage and arching stems of *Danae racemosa* (1) combine beautifully with the richly variegated cream and green foliage of *Euonymus fortunei* 'Silver Queen' (2). A good partnership for shade.

MAHONIAS

The mahonias are named for an Irishman, Bernard M'Mahon, who emigrated to Philadelphia, USA, at the time of the Troubles in Ireland at the end of the 18th century. There he set up a nursery and began to export seed of native American plants. He was frequently given seed of newly discovered plants, which he then raised and distributed. The two American explorers Meriwether Lewis and William Clark, on their epic journey across the centre of the USA to the Pacific Ocean, found *Mahonia aquifolium,* the shiny Oregon grape, growing in the gorges of the Columbia River in 1805 and gave seed to M'Mahon. The Scot David Douglas also noted the plant 20 years later when he too was exploring the north-west Pacific coastal areas.

Mahonia aquifolium bears evergreen, glossy leaves and terminal racemes of rich yellow flowers in spring, followed by blue-black berries. It is a suckering shrub, which will grow almost anywhere. The best selection is *Mahonia aquifolium* 'Apollo' (**1**), which forms a denser shrub with prickle-edged leaves that have shining deep green upper surfaces and red stalks, and produces bright yellow flowers in large clusters. In midwinter the leaves take on mahogany and red tints. Another good selection for winter colour is *Mahonia aquifolium* 'Atropurpurea' (**2**) whose green leaflets become reddish purple in winter, with the low sunshine polishing the leaves. These two forms are seldom more than 1m (3ft) high and about the same across, although they can become straggly; they will happily share space with a golden ivy such as *Hedera helix* 'Amberwaves' and clump-forming perennials such as *Geranium phaeum.*

The attractive hybrid *Mahonia × wagneri* 'Moseri', around 60cm (2ft) high, is unremarkable in flower but the leaves open bronzy red in spring, turn pale green and then shades of coral to deep red in autumn and remain that colour in the winter. It seems unfussy as to soil, but it needs to be grown in sun to get the best effect. *Mahonia × wagneri* 'Pinnacle' (**3**) is often offered as the rare *Mahonia pinnata.* Its bright green leaves are bronze when young and it produces clusters of showy yellow flowers in early spring. For other, especially well-scented mahonias, see page 142.

Viburnum davidii

Viburnum davidii forms a spreading evergreen shrub up to 1.5m (5ft) high, flowering sparsely in late spring, but with the bonus of blue berries if you grow both male and female forms. However, its handsome leaves in a shining dark green, channelled along the veins, are reason enough for growing a solitary plant. It looks good combined with the arching stems of the winter jasmine (*Jasminum nudiflorum*) sprinkling its yellow flowers over the viburnum.

OTHER CLASSIC EVERGREENS *Aucuba japonica* • *Elaeagnus × ebbingei* • *Griselinia littoralis* • *Ligustrum lucidum* •

The evergreen **skimmias** are excellent structural shrubs for winter planting but they also offer colourful flower buds, scented flowers and even berries if you grow both male and female cultivars. Skimmias like shade and are happy in most soils, but the foliage tends to turn a sickly yellow if exposed to too much sun in alkaline soil. They can be lightly trimmed back after flowering in the spring to keep them as neat, mounded shrubs and to encourage more flowers. They form their flower buds in winter and these remain unopened for a very long time, finally developing into conical clusters of scented flowers in spring. The female clones then bear berries, green at first, becoming flushed pink and gradually turning to sealing-wax red and remaining on the plant right through the winter. The birds seem to leave these berries until last, after they have stripped the hollies. The male forms are the most vigorous, with the best trusses of cream flowers.

For year-round perfection the male ***Skimmia japonica* 'Rubella'**🏆 is hard to beat, although it tends to be rather overused as a container plant. ***Skimmia japonica* 'Nymans'**🏆 is a female, with abundant large red berries; ***Skimmia japonica* 'Kew White'** has white berries, and ***Skimmia japonica* 'Fragrans'**🏆 is the one to choose for scent. ***Skimmia × confusa* 'Kew Green'**🏆 is perhaps the finest garden plant, with light green leaves and large clusters of creamy-white fragrant flowers. It is more tolerant of sun than other skimmias. All these grow to form rounded shrubs usually less than 1m (3ft) high.

Skimmia japonica 'Rubella'

Skimmia japonica 'Nymans'

Osmanthus heterophyllus 'Variegatus'

Skimmia × confusa 'Kew Green'

Vinca minor 'Illumination'

Osmanthus heterophyllus is an attractive evergreen with neatly pointed leaves. It is often mistaken for a holly, but the leaves are arranged in pairs, whereas in hollies they alternate along the stems. Most osmanthus grow into small trees or large shrubs but the variegated forms are generally less vigorous. In the autumn the unobtrusive cream flowers give off a powerful perfume. ***Osmanthus heterophyllus* 'Gulftide'**🏆 is a small dense shrub with curiously lobed and twisted leaves. ***Osmanthus heterophyllus* 'Goshiki'** has lime-yellow leaves mottled gold, and bronze-tinted when young (see page 95). In ***Osmanthus heterophyllus* 'Variegatus'**🏆 the leaves (some prickled, some plain) are outlined and splashed with cream, creating a light and airy shrub.

The mat-forming lesser periwinkle, ***Vinca minor***, often makes new growth in winter, providing fresh green leaves that may be plain green or variegated with cream. The usually blue flowers appear sporadically throughout the winter, with a full flush late in the season. ***Vinca minor* 'Atropurpurea'**🏆 has ruby-red flowers and ***Vinca minor* 'Azurea Flore Pleno'** has double flowers in rich purple. ***Vinca minor* 'Illumination'** is a new form, with very bright yellow variegated foliage. Beware the vigorous, creeping ***Vinca major***, which will insinuate itself into everything in sight (in fact, even *Vinca minor* is difficult to eradicate once planted). For a more unusual vinca choose ***Vinca difformis***🏆, which has a rather different habit, with arching, shrub-like growth and palest icy blue flowers.

Conifers

Conifers in a winter garden need to be handled with a light touch. Although they confer a year-round presence, used too freely they can be depressingly monotonous. Just a few add structure to the garden, and if those chosen have coloured leaves, or change or intensify in colour in cold weather, so much the better. Recently the range available has been extended considerably with conifers with blue, gold, cream or variegated foliage, but even more impressive are those that develop deep reddish-bronze tints after the first frosts.

Some of the best silhouettes in winter are provided by the **yews**, in particular the narrow, columnar Irish yews, all of which derive from two plants found on the moors of County Fermanagh in Ireland in 1708. ***Taxus baccata*** **'Fastigiata'**🏆, with a dense, compact column of branches, is indispensable if you want a narrow, dark green pillar, but it does become broader with age, ultimately growing to 5m (16ft) with a spread of 3m (10ft). An even better choice is ***Taxus baccata*** **'Fastigiata Robusta'**, with more tightly packed growth, which is exceedingly slow-growing, reaching only 2m (6ft) in ten years. Equally slow-growing are ***Taxus baccata*** **'Standishii'**🏆, the best golden-leaved Irish yew, and ***Taxus baccata*** **'Ivory Tower'**, which has golden leaves tipped with cream after frost.

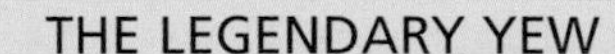

THE LEGENDARY YEW

One of the few native British evergreens, the yew has become inextricably associated with churchyards, where ancient trees hundreds of years old survive. Some say it was a protective tree, planted to drive away evil spirits; other traditions give yew a darker, more sombre side, claiming it absorbed poisonous fumes given off by the dead.

Taxus baccata 'Standishii'

The **cryptomerias** have feathery foliage that often turns bronze and purple in cold weather. ***Cryptomeria japonica*** **Elegans Group** forms a tall, bushy shrub or small tree with delicate ferny juvenile foliage, turning bronze in winter, which it retains throughout its life. ***Cryptomeria japonica*** **'Elegans Compacta'**🏆 is smaller and ideal for the average garden. ***Cryptomeria japonica*** **'Elegans Nana'** has softer,

Cryptomeria japonica Elegans Group

OTHER GOOD CONIFERS *Abies koreana* 'Silberlocke' • *Abies nordmanniana* 'Golden Spreader' •

looser growth and purple winter leaves, and the dwarf ***Cryptomeria japonica*** **'Pygmaea'** is a compact bush with drooping branches, bronzing in winter. These conifers prefer a moist soil and do not do well in exposed sites.

The ***Podocarpus*** hybrids are becoming popular as winter conifers. ***Podocarpus nivalis***, a hardy species from alpine areas in New Zealand, forms a low, spreading mound with crowded stems in olive green, or in bronze in the cultivar ***Podocarpus nivalis*** **'Bronze'**. ***Podocarpus*** **'Young Rusty'** is best in winter when the leaves turn reddish bronze; it is a female and produces scarlet berries when pollinated by a male. ***Podocarpus*** **'Chocolate Box'**, also female, has leaves the colour of dark chocolate when touched by frost.

There are many handsome **pines**, most of which are too tall for the average garden, but the dwarf alpine pines are much easier to accommodate. In some the foliage turns golden yellow in frost, such as ***Pinus mugo*** **'Ophir'**, less than 50cm (20in) tall, or the more spreading

Left: *Taxus baccata* 'Fastigiata' is a wonderful dark green structural conifer.

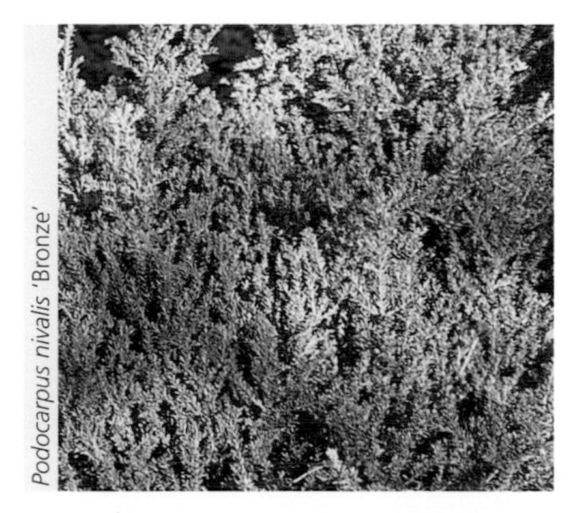

Podocarpus nivalis 'Bronze'

Thuja occidentalis 'Rheingold'

Pinus mugo 'Winter Gold'

Cupressus arizonica var. *glabra* 'Blue Ice'

Pinus mugo **'Winter Gold'**. The dark green ***Pinus densiflora*** **'Alice Verkade'** eventually grows to a height of 1–1.2m (3–4ft), with densely packed leaves, and makes a good rounded bush.

Some green **thujas** deepen to rich gold in winter and of these ***Thuja occidentalis*** **'Rheingold'**♀ is undoubtedly the most popular. In winter the adult foliage becomes dark antique gold with orange tips, turning a fresher yellow-green in summer. It grows into a large, loosely conical shrub but is easily pruned to a manageable shape. The dwarf ***Thuja plicata*** **'Rogersii'** has a conical habit and bronze winter foliage, while ***Thuja plicata*** **'Stoneham Gold'**♀, also conical, has more coppery tones. ***Platycladus orientalis*** **'Aurea Nana'**♀, related to the thujas, grows as a narrow bush with leaves arranged vertically, resembling a candle flame, and it keeps its vivid yellow-green colour all winter (see page 97).

Cupressus arizonica **var.** ***glabra*** **'Blue Ice'**♀ is a striking conifer with leaves in clear blue-grey and small rounded cones, freely borne. It eventually makes a medium-sized conical tree and is at its brightest when caught by the winter sun.

More conifers with coloured foliage are described on pages 96–97, 101 and 107.

CONIFERS FOR GROUND COVER

The spreading ground-cover conifers contribute colour and texture to the lower level of planting. Those with blue foliage such as *Juniperus horizontalis* 'Wiltonii'♀ and *Juniperus horizontalis* 'Bar Harbor' form mats of long, ground-hugging branches clothed in steely blue-grey foliage. Some conifers change colour in winter: the soft green summer foliage of the spreading *Microbiota decussata*♀, for example, turns to bronze-purple in winter (see page 101). *Juniperus* × *pfitzeriana* 'Old Gold'♀ (below) is more compact, with a spread of about 1m (3ft); the branches of bronze-gold foliage hold their colour throughout the year. Ground-cover conifers make a good background for early bulbs and are excellent partners for large-leaved ivies, especially on steep banks.

Architectural plants

Much used by garden designers, vertical accent plants are becoming very popular in modern gardens and they certainly serve as a good counterpoint to low-growing, mound-forming plants. In winter this contrast is more intense and needs to be carefully considered. In the past, plants such as cordylines, phormiums, astelias and yuccas would suffer in colder climates, but the milder conditions of recent years are much more to their liking and they can be grown with confidence in a reasonably sheltered border; they also make excellent container subjects.

The spiky form of *Cordyline australis* grown in a container creates an exotic focal point in the garden in winter.

The hardy **phormiums** come in many colours and they offer evergreen clumps of glossy, sword-like leaves, turning over at the tips as they grow. They are striking when well placed, rising up from low-growing neighbours or planted in gravel, perhaps in a courtyard garden, but they do need full sun. They are harder to use in a mixed border or among shrubs, and the leaves can become damaged and tattered over winter, especially in a windy site. The hardiest of the yellow-leaved varieties is ***Phormium* 'Yellow Wave'**🏆, with soft leaves with drooping tips, reaching 80cm (32in) in height (see page 148). The larger ***Phormium tenax* 'Variegatum'**🏆 provides a more definite arching shape, with its broad leaves, 1m (3ft) long, striped with cream and green. When it comes to purple and red phormiums some of the larger, plain purple forms tend to be the hardiest. ***Phormium tenax* Purpureum Group**🏆 makes a bold plant up to 1.5m (5ft) or more high, with leaves of a dramatic bronzy purple, turning a pleasant pinkish purple in winter.

***Phormium* 'Bronze Baby'** is worth considering for smaller gardens. It forms a compact clump of broad, bronze-purple leaves, 1m (3ft) long, drooping at the tips. ***Phormium* 'Platt's Black'** is a low-growing cultivar, 60cm (2ft) high, in fashionable black, and the colour intensifies as the season progresses, assuming reddish tints in winter. ***Phormium cookianum* 'Flamingo'** comes in rainbow colours of pink, coral, apricot and green; it has compact leaves up to 75cm (30in) long, and is an excellent container plant.

Yuccas are also useful architectural plants, with statuesque flowering spikes in summer, and they are surprisingly hardy. Their fans of sword-like foliage form vertical accents but they can grow quite large. ***Yucca flaccida* 'Golden Sword'**🏆 has soft spring-green leaves with gold centres, and in ***Yucca filamentosa* 'Variegata'**🏆 the edges are ivory white. ***Yucca gloriosa***🏆 has stiffer, upward-pointing leaves, armed with sharp spines, and produces glorious spires of creamy flowers that tower above the foliage. ***Yucca gloriosa* 'Variegata'**🏆 has leaves with pale yellow stripes, fading to buff in winter.

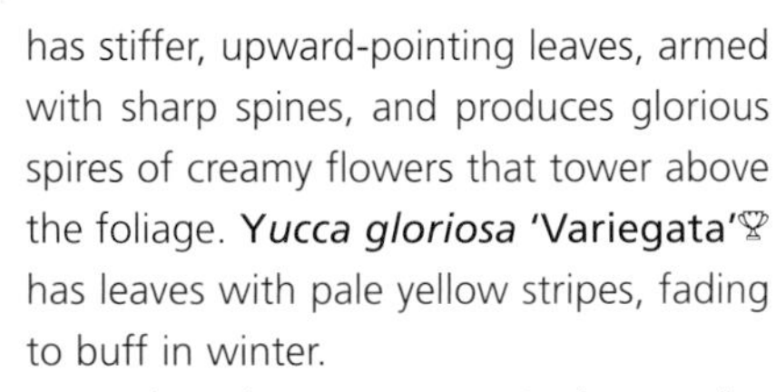

Hardy palms are increasingly popular, especially for pots. The hardiest and most manageable is the dwarf fan palm, ***Chamaerops humilis***🏆, which is native to the Mediterranean. ***Chamaerops humilis* var. *argentea,*** from the Atlas mountains, is a striking silver-blue form (see page 131).

Cordylines grow especially well in warm coastal regions and in sheltered gardens elsewhere and they also make ideal specimen plants in pots. ***Cordyline australis* 'Red Sensation'** has deep burgundy, sword-shaped, leathery leaves spiralling up the main stem, which can reach 2m (6ft) in height. It is able to cope for many years in a large container.

OTHER ARCHITECTURAL PLANTS *Agave americana* • *Astelia nervosa* 'Westland' • *Beschorneria yuccoides* •

Phormium tenax 'Variegatum'

Chamaerops humilis var. *argentea*

Cordyline australis 'Torbay Dazzler'

Phormium 'Bronze Baby'

Cordyline australis 'Torbay Red'

Astelia chathamica

Cordyline australis **'Torbay Red'**🏆 is similar, with deep burgundy-red foliage. ***Cordyline australis*** **'Torbay Dazzler'**🏆, with pale cream striped leaves, is smaller in stature than most cordylines and is a good choice for a pot.

Although not entirely hardy, the fine silver-leaved **astelias** from New Zealand are proving capable of surviving more frost than originally expected. Somewhat surprisingly for silver-leaved plants, they are happy in light shade and respond to a rich soil. Resist the temptation to plant them in a hot, dry spot, where most silver-leaved plants do best. In a shady bed, they will make a dramatic counterpoint to the long silver leaves of *Pulmonaria* 'Diana Clare', or surround them with snowdrops and the white spring flowers of *Anemone nemorosa* 'Vestal'🏆, a semi-double form of wood anemone. ***Astelia chathamica***🏆 has elegant, arching leaves that shine out from a shady spot on a dull winter's day; ***Astelia nervosa*** has narrower leaves with a soft, silky texture. Mulch plants with leaf mould each year to offer some protection and to conserve moisture, but keep the mulch away from the crown. Alternatively astelias can be grown in large containers, where their sword-shaped leaves, sparkling silver above and downy beneath, make a dramatic show. When they outgrow their pots they can be divided in spring and repotted in rich compost.

USING SPIKY PLANTS IN THE GARDEN

Plants with sword-shaped leaves and spiky silhouettes are striking subjects that can be used as focal points in a planting scheme. Their sharp forms contrast dramatically with soft, mounded perennials such as heucheras, and their vertical lines can be used to punctuate carpets of ivy and juniper. Their exotic foliage is shown to advantage against the coloured bare winter stems of dogwood (*Cornus*). Even in the most traditional country gardens they can be used in containers close to the house to provide height and structure in the foreground of the view from the windows.

PROTECTING CORDYLINES IN WINTER

Cordylines can be damaged in winter when water collects in the crown of the plant and then freezes; this can kill the growing tip. In severe weather protect plants by gathering the leaves together and wrapping them gently with a strip of hessian or horticultural fleece. Alternatively the leaves can be gathered and a 'sock' of fleece slipped over the plant.

Butia capitata • *Fascicularia bicolor* • *Phoenix canariensis* • *Sisyrinchium striatum* • *Trachycarpus fortunei* •

Persistent perennials

Most herbaceous (non-woody) perennials shed their leaves at the end of the growing season and die down to ground level, to reappear in the spring, but some retain their leaves right through the winter. An attractive characteristic of some persistent perennials is that, as temperatures drop in the autumn, the foliage becomes tinged with deep red, purple or coral pink. This gives great scope in the winter garden for introducing a contrasting colour among the plain evergreens.

The large leaves of many bergenias develop rich ruby and mahogany tones in cold winter weather.

Bergenias, known as elephant's ears, have large, leathery, spoon-shaped green leaves that in many varieties are transformed in winter, becoming a glowing mahogany with a lighter crimson reverse. They form low clumps up to 30cm (12in) high and, in spring, white, pink or red flowers appear among or just above the leaves. In the past ***Bergenia* × *schmidtii***🏆 gave this genus a bad name with its straggling, brown-blotched green leaves and dull pink flowers intermittently produced from late winter. In recent years newer cultivars have been bred with smaller, neater leaves, better flower colour and the ability to turn shades of ruby red in the winter.

Bergenia 'Bressingham Ruby'

Bergenia 'Eric Smith'

Bergenias that colour well in the winter should be grown in full sun for best effect and they need a well-drained soil; too rich a diet can cause the leaves to collapse into a soggy heap with the arrival of the first frost.

Bergenia* 'Bressingham Ruby'** is an excellent plant, with simple leaves with crinkled edges, in winter becoming dark mahogany on the upper surface, crimson on the reverse. ***Bergenia* 'Eric Smith'** is one of the very best for winter colour, with leaves of polished burgundy. The smaller ***Bergenia* 'Wintermärchen'** (WINTER FAIRY TALES) has narrow, pointed leaves twisting to reveal the brighter reverse, some of the leaves turning scarlet. ***Bergenia stracheyi is also small, with neat, upward-pointing leaves, remaining green, and small sugar-pink flowers in spring. Plant the spring-flowering *Anemone blanda* 'White Splendour'🏆 beside it and *Persicaria microcephala* 'Red Dragon', a sprawling perennial with dark

The foliage of *Epimedium* × *versicolor* 'Neosulphureum' turns rich shades of burgundy, contrasting beautifully with the lime-green flowers of *Helleborus argutifolius* in semi-shade.

red stems and pointed wine-red leaves, to rise up behind it in the summer.

These bergenias are all excellent value in the garden, going on to produce flowers in shades of pink, cerise and cherry red in spring, and their glossy green leaves act as a foil for other flowers in summer.

After a few years the plants can become leggy with lots of bare stems and fewer leaves. At this stage they are best discarded: lift and divide the clump and replant the most vigorous sections to provide new plants.

Epimediums with overwintering leaves are a good choice for woodland conditions. In many cases the leaves turn attractive shades of chestnut, mahogany and polished coppery red in the autumn, bringing unexpected colour at ground level. They mostly form low, slowly spreading clumps of small, leathery, heart-shaped leaves on stiff, wiry, branching stalks; often the leaves are further divided into broad arrow-shaped leaflets. In spring the jewel-like flowers appear, in an enchanting array of colours and intriguing shapes. Epimediums are good space fillers, growing happily in the shade with the stinking hellebore (*Helleborus foetidus*🏆) and the common snowdrop, or associating well with hardy geraniums like *Geranium macrorrhizum* and *Geranium nodosum* and the taller arching stems of Solomon's seal (*Polygonatum*). They can also be planted in more open situations in a border, where their foliage will blend with summer flowers.

Epimedium* × *rubrum🏆 was the result of a very early cross made in Belgium in 1854 (between *Epimedium alpinum* and

Epimedium × *rubrum*

Asarum europaeum

The spiky blue-green foliage of dianthus is here studded with golden yellow winter aconites.

Epimedium grandiflorum🏆). In autumn the dying leaves become a rich reddish brown, lasting well right through the winter. In fact it is hard to find the right time to cut them off in order to see the early spring flowers in crimson and cream, resembling little columbines. The new leaves in spring

Snowdrops and cyclamen peep through the purple-black foliage of *Ophiopogon planiscapus* 'Nigrescens'.

are fresh green edged with bronze and they make pleasant leafy mounds in summer. Honesty is a good companion, the silver skeletons of the seed cases complementing the mahogany epimedium leaves. ***Epimedium* × *perralchicum***🏆 has a creeping rootstock and makes excellent ground cover in dry shade, with glossy evergreen foliage and egg-yolk yellow flowers; the form most generally encountered is **'Wisley'**, which has bigger flowers and brighter foliage. ***Epimedium* × *versicolor* 'Neosulphureum'** (see page 55) has coppery red young foliage, turning burgundy-purple overall in winter, and pale primrose bells in spring. ***Epimedium* × *versicolor* 'Sulphureum'** 🏆, with darker yellow flowers, remains dark green and is a very vigorous spreader, so 'Neosulphureum' is a better choice.

Although remaining green, ***Asarum europaeum*** is a useful carpeter for deep shade, with glossy, bottle-green, kidney-shaped leaves with paler green veining. A northern European plant, it can cope with harsh weather, and the leaves form dense patches close to the ground; they make a good contrast with the smaller leaves of some of the persistent saxifrages, *Saxifraga umbrosa,* and small ferns.

Some persistent perennials come in different colours. **Dianthus** foliage provides a welcome patch of steely blue in winter, while the charcoal-purple grass-like mounds of ***Ophiopogon planiscapus* 'Nigrescens'**🏆 make a wonderful background for pure white snowdrops or glowing pink cyclamen.

Many of the **euphorbias** have handsome overwintering leaves in various colours: glaucous blue, dark purple or variegated in cream and green. The two Mediterranean forms ***Euphorbia characias*** and ***Euphorbia characias* subsp. *wulfenii***🏆 have pewter-grey leaves overlapping in a swirl around the stems and are almost shrub-like in stature, growing 1–1.2m (3–4ft) high (see page 107). In winter the newest leaves are tinged with red and the stems too turn maroon. The flowering stems gradually elongate over the winter, bending their necks like swans, only to lift their heads high when the spectacular flowering bracts finally open. These stems are biennial, growing one year to flower and die the next, and the plants can be kept in good shape by cutting the spent flowering stems at ground level in summer to allow room for new growth for the winter.

These big euphorbias look well growing as specimen perennials surrounded by low-growing plants. *Veronica peduncularis* 'Georgia Blue', with purple-tinted foliage and bright blue flowers (see page 33), can be allowed to weave its way beneath a particularly fine yellow flowerer such as ***Euphorbia characias*** **subsp.** ***wulfenii*** **'Lambrook Gold'**♀.

Euphorbia × *martini*

Euphorbia characias

Euphorbia characias SILVER SWAN ('Wilcott')

Euphorbia amygdaloides 'Purpurea'

Other smaller euphorbias have almost sea-blue leaves: ***Euphorbia myrsinites***♀, for example, with spiralling stems at ground level, and ***Euphorbia*** **'Blue Haze'**, growing up to about 60cm (2ft) high. Contrasting colours come in ***Euphorbia characias*** SILVER SWAN (**'Wilcott'**), with sage-green leaves narrowly outlined in ivory, and in ***Euphorbia amygdaloides*** **'Purpurea'**, which has grey-green leaves, tinged purple on the new spring growth, on dark red stems. ***Euphorbia*** BLACKBIRD (**'Nothowlee'**) is a low hummock of a plant with grape-purple, almost black, foliage when grown in sun (see page 130).

Euphorbia × ***martini***♀, a natural hybrid between *Euphorbia characias* and *Euphorbia amygdaloides*, makes a pleasant small bush of olive green with hints of

(continued on page 60)

HANDLING EUPHORBIAS

The white sap of euphorbias is a painful irritant, so always wear gloves when handling these plants and be very careful not to rub your eyes.

Euphorbia myrsinites

HEUCHERAS

The hardy leaves of purple heucheras look particularly beautiful when etched by frost.

The heucheras are members of the Saxifragaceae family and they are all native to the USA, with the numerous species distributed throughout the entire country. Not many were considered garden worthy until the last 20 years, when breeders started deliberately hybridizing the species to produce a huge variety of different leaf colours and shapes – and these have proved winners in gardens in the 21st century.

Heucheras, and the related heucherellas, tiarellas and tellimas (see page 60), have enjoyed a marked rise in popularity in recent years and are set to rival hostas in universal appeal. In the past they were grown mainly for their pretty spires of small red flowers and were known as coral bells. They are now grown as foliage plants as well as for their flowers, and their attractively coloured leaves persist through the winter.

The first break came in the 1980s when a distinctly purple-leaved variant of *Heuchera micrantha* var. *diversifolia* (although the parent is now considered to be *villosa*) occurred in a batch of seedlings sent to the Botanic Gardens at Kew from the USA. This selection was planted out in the Queen's Palace Garden at Kew and named 'Palace Purple' and was the forerunner of a flood of dark-foliaged plants. Several noted American nurserymen then began to develop heucheras in colours ranging through purple, chocolate, brown, bronze, red and pewter. In many the leaves are veined and splashed with silver, and they may be ruffled, curled or crimped at the edges. More recently, leaves in tan, orange and caramel have appeared, and now chartreuse green has been added to the list.

Not all these new heucheras are proving to be long-lived garden plants but some are excellent and are here to stay. Although heuchera leaves persist over winter, to be replaced by new growth from the centre of the plant in spring, some do become decidedly tatty and cannot be rated as winter-foliage plants; 'Palace Purple' falls into this category. However, some of the more vigorous hybrids with genes from winter-tolerant species are invaluable in the winter border, being robust plants able to cope with cold weather – although persistent wind may damage some of the leaves.

As the cold intensifies, the silvery veins stand out on the purple-leaved forms, as do the wine-red veins on some of the green- and silver-leaved plants. One that is highly recommended is *Heuchera* 'Stormy Seas' (**1**), hardly flinching in even the stormiest weather, with purple and pewter leaves veined with silver. *Heuchera* 'Beauty Colour' (**2**) is also a winner, turning almost black with silver highlights in cold weather; it needs to be grown with a blue-leaved dianthus or copper-leaved sedge for contrast. *Heuchera* 'Chocolate Ruffles', with strongly ruffled, brownish-purple leaves showing off the maroon undersides, and the older variety 'Plum Pudding' (**3**), in a mixture of purple, crimson and silvery grey, show up well in the winter border, as does the silver *Heuchera* 'Venus', while *Heuchera americana* 'Ring of Fire' gains an outline of coral round each leaf at the onset of frosty weather. Of the green-leaved forms, *Heuchera* 'Green Spice' (**4**) is stunning, with well-shaped, mint-green leaves shot with burgundy-red veins in winter. The newer cultivar *Heuchera* 'Marmalade' has ruffled leaves in a mixture of olive green, tan and orange, with the undersides showing rhubarb pink – quite eye-catching on a grey winter day – and *Heuchera* CRÈME BRÛLÉ ('Tnheu041') (**5**) adds tones of amber and ginger. *Heuchera* KEY LIME PIE ('Tnheu042') (**6**) and 'Lime Rickey' offer the latest colour break in heucheras, with leaves in bright lime green, but these varieties have yet to be tested for winter performance.

Heucheras vary in stature but most form a rounded mound of leaves 30cm (12in) high and the same across. Fine stems carry tiny flowers way above the foliage, often at a height of 60cm (2ft) or more. The varieties with amber- and lime-coloured leaves tend to make smaller and more compact plants.

There is one heuchera, bred in England, that should be more widely available: it is an excellent winter-foliage plant, with bronze, red and purple leaves, and long-lasting, pale pink flowers as well. This is *Heuchera* 'David' (**7**), which develops into a vigorous spreading mound, happy in full sun, and it can be left to itself for many years. It was bred by Mary Ramsdale in 1994 when she held the National Heuchera Collection.

GROWING HEUCHERAS

Heucheras are content with many positions in the garden, tolerating both full sun and dry shade, but are probably less suited to damp, heavy clay and winter wet; the purple-leaved forms are best in sun as their colour can fade in shade. In the wild many of them are woodland plants, forming the understorey beneath tall trees, where they grow into weed-smothering ground cover. They are also fantastic container plants with foliage colours to combine with almost anything.

CARE OF HEUCHERAS

These tough plants need a tidy up in early spring. Cut back some of the damaged top growth, pulling out old leaves and flower stems, and the plants will renew themselves from the centre.

Heucheras are very susceptible to vine weevil. The cream-coloured larvae feed on the roots during winter causing serious damage to the plants. To prevent this, use a chemical drench or biological control in early autumn and again in early spring.

TIARELLAS, HEUCHERELLAS AND TELLIMAS

There are two North American species of *Tiarella* (commonly known as foam flower) that are most often grown in gardens and these have been hybridized to produce some good new plants. They form low clumping mounds of persistent foliage and some turn to shades of bronze and brown with red veins in winter. *Tiarella cordifolia* 'Glossy' has large, bright green leaves, while *Tiarella wherryi* 'Bronze Beauty' has red-stained green leaves, which darken to russet in the winter. *Tiarella* 'Skid's Variegated' is an amazing plant, with apple-green leaves speckled all over with cream, turning coral and pink in winter (**1**), and it is more vigorous than other variegated forms. The flowers of tiarellas appear as candle-like buds among the leaves, opening to tiny starry flowers on stems rising just above the foliage; they open in succession from late winter right through to early summer. Tiarellas are happiest in light shade, although they will tolerate a sunny site.

Heucherellas are hybrids between heucheras (see pages 58–59) and tiarellas and are generally known as foamy bells. Their flowers are sterile and do not set seed, so plants continue to bloom for a long time, with a profusion of tiny flowers on upright stems, usually 30cm (12in) or more in height. They are probably more suited to woodland conditions than to full sun in the summer. Most are not good winter foliage plants, although × *Heucherella* 'Quicksilver' (**2**) is an exception. Drought tolerant and a vigorous grower, it has silvery metallic leaves, purple on the reverse. In winter the foliage turns mahogany red with pewter marbling. Pink buds open to white flowers on delicate stems in early summer.

There is one superb tellima, *Tellima grandiflora* Rubra Group, which is indispensable in the winter garden. It forms a weed-smothering clump of prettily shaped, almost circular leaves with scalloped edges, which turn a fiery coral red with the first frost. In shade they are more maroon-red. Tiny greenish-white bells dangle from 60cm (2ft) stems in early summer and need to be removed when withered lest they spoil the winter effect. The clumps are not too vigorous for snowdrops to push through, and planted alongside the marbled leaves of *Arum italicum* subsp. *italicum* 'Marmoratum'♀ will form an eye-catching group that will entice you into the garden in winter (see opposite). There is a newer selection called *Tellima grandiflora* 'Forest Frost' (**3**), which has leaves turning charcoal splashed with silver in winter. Tellimas will grow in sun but are at their best in light shade.

PLANTING PARTNERS

The deep red-purple flowers of *Erica carnea* 'Myretoun Ruby'♀ make a wonderful contrast with the frosty white flowers and bright foliage of *Arabis procurrens* 'Variegata'♀.

purple, good in winter and useful in shade, although it will also grow in sun.

Evergreen variegated foliage is effective, too. ***Arabis procurrens*** **'Variegata'**♀ is a carpeting plant for the front of the border, with glossy little rosettes of linear leaves, almost entirely cream with a central splash of mint-green, pink-tinged in cold weather. By late winter little sprays of white flowers begin to appear, rising above the foliage

A leafy tapestry of tellimas and arums embroidered with snowdrops lights up a shady corner of the garden at White Windows in Longparish, Hampshire.

Origanum vulgare 'Polyphant'

Iris foetidissima 'Variegata'

Lamium maculatum 'White Nancy'

on purple stems. In time it will form an evergreen cushion through which can thrust bulbs such as small irises and crocuses. Like many variegated plants, this arabis tends to produce plain green rosettes among the patterned ones. Remove these as they appear, or they will gradually take over.

Marjoram is a fine winter evergreen, and the variegated form ***Origanum vulgare*** **'Polyphant'** makes a tight mound of cream and green. ***Iris foetidissima***🏆, with evergreen sword-shaped leaves in dark green, or green and cream in **'Variegata'**🏆, provides strong leaf contrast in a shady setting.

Forms of the deadnettle ***Lamium maculatum*** can persist well through the winter with leaves that are almost entirely silver. If cut back hard in early autumn the plants will remain as neat mounds, to spread and flower in spring and summer. ***Lamium maculatum*** **'White Nancy'**🏆 is a lovely combination of silver leaves and white flowers. (See also page 94.)

All these evergreen perennials planted in combination with shrubs, trees and early-flowering bulbs provide a rich tapestry of shape, colour and form throughout the winter months.

WINTER FLOWERS

Winter flowers are precious, bringing colour and scent at a time when they will be most appreciated. The simple snowdrop shines out in purest white and the little winter aconite is ever cheerful in bright yellow. Crocuses form prettily patterned carpets under trees, while sturdy hellebores display a tapestry of subtle colours from palest cream to inky purple. The winter jasmine has showy flowers on bare stems, and the delicate blossom on the winter cherries sparkles in the watery sunshine.

RIGHT: *Crocus tommasinianus* pushing through fallen leaves.

Bulbs

Winter-flowering shrubs add form and structure to the garden but it is the dwarf bulbs providing colour under and around them that tempt the gardener to venture outside on a cold day. The winter aconites are always among the first to appear, bringing flashes of yellow to rival the winter jasmine and witch hazels. Then come the snowdrops (although some of these can be in flower before Christmas) and the cyclamen, and then a rush of small irises, the first crocuses and the early-flowering narcissus.

Eranthis hyemalis 'Guinea Gold'

The winter aconite, *Eranthis hyemalis*, produces cheery yellow flowers from mid- to late winter. It thrives under the shade of trees or large shrubs.

The winter aconites, ***Eranthis hyemalis***♀, can be relied on to start flowering in midwinter. They are dwarf members of the buttercup family bearing bright yellow, globular flowers fringed with a ruff of narrow green leaflets. The small tubers are sometimes difficult to get established but when they are happy they will seed around abundantly, particularly into gravel; they are best increased by being moved 'in the green' just after they have flowered. Preferring to grow in some shade, they are quite impervious to bad weather and can cope with snow and frost, perking up undamaged when the snow melts. There is a later-flowering hybrid, ***Eranthis hyemalis* 'Guinea Gold'**♀, with brighter, golden yellow flowers on taller stems and leaves that are tinged with bronze. This does not appear to set seed and is slow to increase vegetatively, but it is a treasure to grow among special snowdrops.

It is possible to have hardy **cyclamen** in flower right through the autumn and winter if you plant the two species ***Cyclamen hederifolium***🏆 and ***Cyclamen coum***🏆. The former comes into flower in early autumn, on bare stems before the leaves. These flowers have swept-back petals, often with a contrasting dark spot at the base, and are usually pure white or cerise-pink, but deep magenta forms do exist. They seem happiest in dry shade and will comfortably occupy that difficult spot beneath a conifer or at the bottom of a hedge. Here they will multiply, and ants will move the seeds around as they are attracted to the sugary substance that coats them. The patterned leaves, resembling those of ivy (*Hedera* – hence this cyclamen's name *hederifolium*), appear after the flowers have faded and are so marbled and swirled with silver and green that no two plants are identical. They form an intricately patterned carpet that can be interplanted with winter aconites so that the continuity of flower is kept going until *Cyclamen coum* appears in midwinter.

Cyclamen coum seems to prefer more moisture in the soil but can take a reasonable amount of sun in winter. The rounded leaves have the texture of polished leather and they vary greatly in their markings, from two shades of green to silvered edges, pewter centres and almost

Below: Autumn-flowering *Cyclamen hederifolium* has ivy-shaped leaves patterned in silver, sage green and grey. They make a wonderful partner for winter-flowering bulbs such as snowdrops and aconites.

completely silver-grey: ***Cyclamen coum* Pewter Group**🏆 and **Silver Group** are good selections. The swept-back flowers have a twist in the petals, looking rather like a propeller, and come in vivid reds, pinks and snow white, all with a purple stain at the base. Good companions are the black grass-like *Ophiopogon planiscapus* 'Nigrescens'🏆 (see page 56) or purple-leaved heucheras. These cyclamen are long-lived and will spread their seed freely. The two species are best kept apart as the stronger-growing *Cyclamen hederifolium* could crowd out *Cyclamen coum*. If grown

Narcissus 'Rijnveld's Early Sensation' is a useful yellow daffodil flowering from midwinter. It adds colour alongside dark evergreens and brightens the border before herbaceous perennials take over in spring.

Narcissus 'Tête-à-tête'

in containers, they are best kept pot-bound in order to flower well.

It is surprising that ***Narcissus* 'Rijnveld's Early Sensation'**♀ is not more widely planted since its bright yellow trumpets start the daffodil season off in fine style. It has been around for a long time but only recently has it become more readily available. The substantial egg-yolk yellow cups, on sturdy stems up to 40cm (15in) high, can be in flower in midwinter, making a striking companion for the purple hellebore, *Helleborus orientalis* Early Purple Group. Another very early daffodil is ***Narcissus* 'Cedric Morris'**, with perfectly proportioned flowers in lemon-yellow with a frilled trumpet, which often appears by midwinter and continues in flower late into the season. It grows only 25cm (10in) high, with narrow leaves, and it is quite able to cope with freezing weather. This daffodil is slow to increase and needs to be planted where it will not be obscured by larger bulbs later in the winter. ***Narcissus* 'Tête-à-tête'**♀, with small golden yellow flowers, increases well and starts to flower in late winter, as does the pretty ***Narcissus* 'February Gold'**♀.

The first dwarf **irises** start spearing through the ground in midwinter and because so little else is in flower at this time of year, these little gems can be properly appreciated, even to the extent of getting down on hands and knees to catch their elusive scent. In the full thrust of growth in summer such small plants could easily be overlooked. Those most often grown are ***Iris histrioides*** and ***Iris reticulata***♀ and their hybrids, which are reliably early-flowering and remarkably robust in even fierce weather. One of the most unusual is ***Iris* 'Katharine Hodgkin'**♀, a Reticulata iris, with relatively large flowers on short stems, in an ethereal mixture of pale grey and softest lilac with a glint of yellow. Some consider this a rather melancholy plant, but the broad front petal, or fall, has eye-catching purple stripes around the edge and matching spots overlaying the yellow centre, showing up well on a dark day. This is a surprisingly vigorous plant

GROWING DWARF IRISES

In order to get dwarf irises to flower in following years they need to be planted at least 15cm (6in) down, and they prefer chalky or limy soil that gets a summer baking. The fat buds start to push through the ground in midwinter, waiting for some warmer weather before bursting into flower; they are perfectly hardy and don't mind being covered in snow. After flowering, the leaves will elongate and need to be left to die down naturally, so grow them among summer plants with insubstantial foliage.

Iris reticulata

Iris 'Harmony'

and increases freely. Other crosses made with *Iris reticulata* come in stronger blues and purples: ***Iris* 'George'**♕ produces charming flowers in the richest purple with almost black falls; ***Iris* 'Harmony'** and **'Joyce'** have blue flowers, and **'Pauline'** is reddish violet.

The charming small-flowered **crocuses** that flower early in the year are eagerly awaited. Among the very earliest are ***Crocus sieberi***♕, from Greece, with lavender flowers and ***Crocus sieberi* 'Albus'** (formerly 'Bowles' White'), with pure white globular flowers with an orange throat, blooming for a long time in mid- and late winter and giving off a delicious soft perfume. These crocuses like a poor soil and good drainage and can be grown among lightly spreading mats of alpine plants, where they will receive a good baking in the summer. Next to flower are the affectionately named 'tommies', ***Crocus tommasinianus***♕, easily established in gardens, where they often spread with enormous enthusiasm; impervious to bad weather, their slender pointed buds can push up even through snow. It is probably best to plant them in grass or a wild area where they will form a shimmering lavender carpet; an old apple tree is a perfect companion. ***Crocus tommasinianus* 'Whitewell Purple'** is a

Crocus sieberi 'Albus'

Crocus tommasinianus 'Whitewell Purple'

Crocus tommasinianus

darker purple form, which is even better than the species. ***Crocus tommasinianus* 'Ruby Giant'** has more strongly coloured flowers of reddish purple. It spreads well and looks good when naturalized under birch trees.

Some of the early ***Crocus chrysanthus* hybrids** have flowers like slim candles in subtle colour combinations; when happy in a sunny bed, they will multiply well but never become a problem. E.A. Bowles bred many crocuses in his famous garden at Myddelton House in north London, and his ***Crocus chrysanthus* 'Snow Bunting'**♕, with pure white petals with a purple base and golden throat, is still one of the best.

SHOWY EARLY CROCUSES

Some early crocus hybrids have exotically marked blooms, making them showy and flamboyant despite their diminutive size. They are useful in pots and to add clumps of welcome colour at the front of the border.

Crocus chrysanthus 'Ladykiller'♕ Striking blooms are cream inside with deep purple markings on the outside.

Crocus sieberi subsp. *sublimis* 'Tricolor' ♕ Petals are bright lilac at the tips, white in the centre and yellow at the base, giving a banded appearance to the open flowers.

Crocus chrysanthus 'Gipsy Girl' (below) Butter-yellow blooms are marked and striped with dark purple on the outside.

Crocus tommasinianus 'Ruby Giant'

All these small varieties grow to about 8cm (3in) high, and the leaves fade away quite quickly in the spring. Unfortunately birds can be a menace in some gardens, nipping off the petals for no obvious reason. They tend to target yellow flowers and can also attack early primroses and even forsythia, but later in the season they are usually too busy nest building to destroy other flowers.

The shining yellow celandines, ***Ranunculus ficaria,*** can become a bit of a nuisance, popping up all over the garden, but ***Ranunculus ficaria* 'Brazen Hussy'** has dark chocolate-brown polished leaves, which show off the vivid yellow flowers and provide a pleasant foliage colour for this time of year. Although it too spreads quite freely, it soon disappears below ground and can be covered with the leaves of small hostas.

The pale green, grassy leaves of ***Ipheion uniflorum*** are appearing increasingly early, slowly building up into a low mound, with star-shaped flowers usually of icy blue with darker veins. ***Ipheion uniflorum* 'Wisley Blue'**♕ has clear light blue flowers, and **'Charlotte Bishop'** is pink; there is also a handsome, later-flowering white form, with broader, glaucous foliage, called ***Ipheion* 'Alberto Costillo'**.

Ranunculus ficaria 'Brazen Hussy'

MORE BULBS FOR WINTER AND EARLY SPRING *Chionodoxa sardensis* • *Corydalis malkensis* •

FORCED HYACINTHS

Forced hyacinths are often grown in bowls indoors for their wonderfully scented flowers, but when these have finished the bulbs are usually discarded. If you plant them outside instead, they will settle in and continue to bloom in later years, but with much smaller flowerheads, with the petals not so tightly packed. Apart from the usual blue, white and pink forms, there are some beautiful very dark purple ones, delightful when grown in combination with a silver lamium in a sunny spot.

Ipheion uniflorum 'Wisley Blue'

Ipheion 'Alberto Castillo'

Scilla miscatschenkoana

For those impatient for the bulb season to start, there is a snowdrop that flowers in early winter; this is ***Galanthus reginae-olgae***, with delightful slim white bells appearing before the leaves. ***Crocus goulimyi***♀, with pale lilac or white flowers, also comes into bloom extremely early. Both these bulbs survive outside and will slowly increase. At the other end of the season there is the little brilliant blue ***Chionodoxa luciliae***♀, known as glory-of-the-snow in its native Turkey and flowering when the snow melts in the mountains; in Britain it usually comes into flower in early spring. It seeds about enthusiastically, as does the similar but slightly earlier ***Scilla mischtschenkoana***♀, which has ice-blue veined flowers just above emerald leaves.

Chionodoxa luciliae

Crocus etruscus subsp. *flavus* • *Iris* 'J.S. Dijt' • *Leucojum vernum* • *Narcissus bulbocodium* • *Scilla bifolia* • *Tulipa turkestanica* •

Snowdrops naturalize freely in the light shade of trees. They enjoy a soil that has plenty of organic matter and is reasonably moist in the winter.

SNOWDROPS

A snowdrop seems to epitomize the perfect flower for winter: shining white against the sombre evergreens and tough enough to survive the icy weather, it pops up in unexpected places, with the first opening bud heralding the start of a new year in the garden.

Although it is not certain if *Galanthus nivalis*♡ is a British native, it has been growing for centuries in churchyards, where it may have been introduced by the early monks. Snowdrops became known as 'fair maids of February' and were associated with Candlemas, the festival of the purification of the Virgin Mary held on 2 February, as their pristine white flowers seemed a perfect symbol of purity. In the herbals of the 17th century they were called 'bulbous white violets' and it was only in the next century that snowdrop became the accepted name. Since then snowdrops have risen rapidly in popularity, with enthusiasts in their hundreds scouring the country looking at snowdrop collections, dropping on their knees to inspect the smallest differences in green markings on the white petals and paying large sums of money for the very rare cultivars.

In the wild, snowdrops prefer shady slopes and thrive on a limestone soil. They are easy to accommodate in the winter garden, planted either beneath trees or in shady borders. They associate particularly well with hellebores and are happy with the same conditions of light and moisture in their growing season in late winter, and shade in the summer. Special snowdrops can be planted beside herbaceous geraniums such as *Geranium wallichianum* 'Buxton's Variety'♡, with a central core of growth and long flowering stems covering the ground

OTHER GOOD SNOWDROPS *Galanthus* 'Dionysus' • *Galanthus* 'John Gray' • *Galanthus nivalis* Scharlockii Group •

in summer. When this is cut back in the autumn, the snowdrops will have plenty of space to flower before the geranium appears again, but they must be well labelled to avoid something else being planted on top of them. The base of a cool wall, among ferns and small hostas, is another good spot.

The common snowdrop, *Galanthus nivalis*♕ (**1**), and the charming double form with frilly, green-edged petticoats, *Galanthus nivalis* f. *pleniflorus* 'Flore Pleno'♕ (**2**), are easy to grow and quickly build up into substantial clumps. If divided and spread around just after flowering, while they are still in leaf, they will soon provide snow-white patches beneath winter trees.

There are numerous improved hybrids and one of the first to flower is *Galanthus* 'Atkinsii'♕ (**3**), a tall snowdrop with simple hanging bells on long stems. The vigorous, pleasantly scented *Galanthus* 'S. Arnott'♕ (**4**) is a great favourite, with relatively large flowers, and *Galanthus* 'Magnet'♕ (**5**) is another strong-grower with delicate drops on long arching stems. The beautifully marked *Galanthus* 'Desdemona' is a large double and derives from *Galanthus plicatus*♕ (**6**), a species that has striking leaves with a central pleat and good-sized flowers, which has been used in the hybridizing of some excellent snowdrops. For contrasting foliage colour choose *Galanthus elwesii*♕ (**7**) or one of its cultivars, with large leaves in glaucous blue; these need more sun to perform well. Coming into flower late in the season is *Galanthus ikariae* (**8**), a very different snowdrop, with broad emerald-green leaves without a hint of blue and large flowers on short stalks. All these have green markings on the inner petals but a few, such as *Galanthus plicatus* 'Trym', have markings on the outside, and others have yellow markings instead of green, such as *Galanthus plicatus* 'Wendy's Gold'. These special snowdrops can be hard to find, but to a galanthophile (literally a lover of snowdrops) the hunt will be exciting and well worth the effort involved.

Galanthus nivalis 'Viridapice' • *Galanthus plicatus* 'Augustus' • *Galanthus* 'Straffan' • *Galanthus woronowii* •

Flowering perennials

Herbaceous perennials that bloom in the depths of winter are very scarce, so the queen of winter flowers, the hellebore, reigns supreme in this season: what other plant is in flower from midwinter to early spring, giving us warm, glowing colour in the winter gloom?

This bed of *Helleborus × hybridus* is glorious in winter when the plants are in bloom. In summer an assortment of daylilies (*Hemerocallis*) planted between them gives a luxuriant display of flowers and foliage.

HELLEBORES

The increasingly popular hellebore must be the first choice for a winter flower, whether you choose the instantly recognized Christmas rose, ***Helleborus niger*** (see page 75), or some of the richly coloured Lenten roses. These are now known as ***Helleborus × hybridus*** and have been bred from the wild species that occur in the eastern Mediterranean, from northern Italy east to the Balkans and across into Turkey. Species such as *Helleborus orientalis, torquatus, purpurascens* and *odorus* have been crossed to give us the rich range of colours that is available today: muted old tapestry shades from cream to pink, rich red, grape purple and deep burgundy, many freckled with spots on the inside of the pendent flower cups or veined in a contrasting colour. There are singles with rounded petals and outward-facing flowers, singles with pointed petals, doubles and semi-doubles. Many have contrasting dark nectaries, which are in fact the true flowers, nestling at the base of the cup; what appear to be the richly coloured petals are actually

IN SNOW AND FROST

Hellebore stems collapse after a fall of snow or in a frost, but will quickly recover once the temperature rises.

modified leaves, known as sepals. When the nectaries have been fertilized they shrivel and fall off, but the sepals continue to show colour, which is why these plants appear to flower for such a long time.

These hellebores are slow to increase by division and are not very suitable for micropropagation, which means that named cultivars are not often available. However, seed-raised strains are very satisfactory, and a good specialist nursery will provide excellent plants of a uniform colour strain. When grown in the garden they are prolific seeders and it is best to remove the ripening seed pods if you do not want choice plants to be overwhelmed by the seedlings, which will not be the same as the parent plant. Left to themselves, they will revert to a greenish cream or a muddy purple.

The hybrid hellebores are happy growing in shade, in humus-rich, neutral to alkaline soil, and they are very long-lasting plants, resenting disturbance and best left to build up into substantial clumps, sometimes with 40 or more flowers. With heights varying from 20 to 40cm (8–16in), they can be tucked away beneath deciduous shrubs, where they will show up well and will not be taking up premium space for summer flowers in the front of the borders. They could also be planted on a shady bank, where it is easy to lift up the flowers in order to see the intricate speckling within. Alternatively, they can be grown together in a special bed where the full range of colours can be appreciated. Annuals such as the fragrant tobacco plant *Nicotiana affinis* or hybrids of the biennial foxglove *Digitalis purpurea* can be grown among them, if the bed is not too shady, for flowers in the summer.

Hellebores are fine resilient plants, happily bowing to the ground under snow, frost or heavy rain and springing upright again in more benign weather. By the end of the autumn their old leaves are becoming tatty and often marred with brown spots. It is best to cut these leaves

A SELECTION OF HELLEBORE HYBRIDS

The colours seen in *Helleborus* × *hybridus* range from pure white through pale yellow and soft pink to deep red, dusky mauve and the darkest of velvety purples.

Hellebore flowers wilt and fade when stems are cut and put in water. Pick off the heads and float the flowers on water or set them on damp moss to see them at their best.

right off before the new flower stems appear rather than leaving them to spread disease. At the same time apply a mulch of garden compost around the plant roots.

There are also hellebore species that have green flowers. ***Helleborus argutifolius***♀ comes from Corsica (and, helpfully, used to be called *Helleborus corsicus*) and is an unusual member of the hellebore family in that it prefers full sun. It grows on open hillsides in Corsica, where it cascades down the slopes. It is a large plant for a hellebore, almost shrub-like in well-grown specimens. The handsome saw-edged leaves are divided into three spiny leaflets, and from midwinter onwards the stems

Helleborus argutifolius

PLANTING PARTNERS FOR HELLEBORES

Hybrid hellebores lack foliage when the plants are in flower (the old leaves should be removed prior to flowering), so planting partners that supply a leafy background are particularly welcome. Evergreen ferns such as *Asplenium scolopendrium*♀, the hart's tongue fern, and *Dryopteris erythrosora*♀, the Japanese shield fern, enjoy similar growing conditions. Dwarf evergreen shrubs such as the Christmas box, *Sarcococca confusa*♀, and *Euonymus fortunei* 'Emerald 'n' Gold'♀ (above) will provide some shelter on more exposed sites if the hellebores are planted between them. Grasses such as *Carex comans* and *Carex buchananii*♀ add lighter texture and work well when planted with the darker shades of hellebore flowers. *Asarum europaeum* forms a carpet of rounded shiny evergreen leaves, through which the hellebore flowers will rise.

THE CHRISTMAS ROSE (HELLEBORUS NIGER)

Snowdrops bloom at the same time as the Christmas rose, *Helleborus niger*. Both plants have long been grown in cottage gardens.

Helleborus niger♀, the Christmas rose, is neither a rose nor very often in bloom at Christmas time, but this beautiful flower is the best known member of the hellebores and has a fascinating history. Native to mountainous areas in northern Italy, southern Germany and east to Slovenia and Croatia, it probably came to Britain with the Romans and was at first used only as a medicinal plant. The *niger* part of the name, meaning black, refers to the black or dark brown roots, which are very poisonous, inducing violent vomiting and convulsions, but carefully controlled doses were considered effective in the curing of lunacy and manic depression. In Gerard's herbal in the 17th century, it was recommended as a purgative for 'mad or furious men'.

The Christmas rose usually bears large, saucer-shaped white flowers with prominent yellow stamens, and in some forms the buds are flushed with pink and the flowers fade to a soft rose-pink. The leathery leaves, divided into several segments, can often hide the flowers in poor forms, but selected seed-raised strains are available that hold the flowers well above the foliage: *Helleborus niger* Blackthorn Group and *Helleborus niger* 'White Magic' are recommended. There are also cultivars that have been selected for early flowering and it is possible to have blossoms to pick on Christmas Day, although the plant usually comes into flower a little later, in midwinter.

Helleborus niger is often difficult to grow well, although in some old cottage gardens there are long-established clumps that flourish with little attention. It seems to prefer a slightly alkaline soil, rich in humus, and not too much shade. The plants are long-lived and resent disturbance; they are easily raised from seed, sown in summer. The leaves can become very tattered and marked with black spots and it is best to remove most of the old foliage from the centre of the plant before it comes into flower.

There are some good double forms of *Helleborus niger* coming onto the market and the species readily crosses with the Corsican hellebore, *Helleborus argutifolius*♀, to produce *Helleborus* × *nigercors*♀, with even larger, pure white flowers and more blue-green leaves with fewer divisions. Exciting new developments are taking place with the introduction of the elusive *Helleborus thibetanus* (**1**) into cultivation and this has been crossed with *Helleborus niger* to produce *Helleborus* 'Pink Ice', with sugar-pink flowers and marbled foliage. Further crosses have been made with *Helleborus vesicarius,* resulting in *Helleborus* 'Briar Rose' (**2**), which has smaller flowers with dark red edges.

The other well-known green-flowered species, ***Helleborus foetidus***♕ (stinking hellebore), is better suited to growing in woodland or wilder areas, where it can seed about (see page 123). ***Helleborus foetidus* Wester Flisk Group** has exceptionally good dark green foliage and lime-green flowers on red stems. It is more compact and rounded than the species.

Although the species hellebores are not as flamboyant as their hybrid cousins, they do have a quiet appeal. One of the smallest is ***Helleborus dumetorum,*** with luminous jade-green flowers. The much-divided leaves develop as the flowers are fading; the slender stems of the woodruff *Asperula aristata* subsp. *scabra*, with

Helleborus × *sternii* Blackthorn Group

Above: *Helleborus foetidus* Wester Flisk Group provides a striking contrast to conifers and heathers with its bold, dark leaves and green flowers on red stems.

Right: The bright green foliage of *Helleborus dumetorum* sets off beautifully the pearl-white drops of *Galanthus* 'S. Arnott'.

are topped with multi-headed clusters of little cups of palest green, almost cream, each with a starburst of stamens filling the centre. Even before the flowers appear the handsome foliage, caught by an early hoar frost, is a striking sight. Grow it in front of dogwoods and willows with coloured stems, or even at the base of a beech hedge – the chestnut-brown hedge and the apple-green flowers make an attractive grouping.

gypsophila-like white flowers, would be happy weaving through them in summer.

In the 1950s Sir Frederick Stern was the first to make the cross between the Corsican hellebore and the tender, pink-flowered *Helleborus lividus*🏆, and this cross became known as ***Helleborus* × *sternii***. In good selections it has pale marbling on the leaves, red stems and pink tones in the flowers, resembling the Christmas rose in shape, on neat plants about 30cm (12in) high. ***Helleborus* × *sternii* Blackthorn Group**🏆 is an excellent selection with attractively silver-veined green foliage and greenish-pink flowers. Some of the crosses are taller, nearer to their Corsican relative, and without the pink tones; ***Helleborus***

Helleborus × ericsmithii

× *sternii* 'Boughton Beauty' is the best of these. There is even a three-way cross involving the genes of *Helleborus niger* as well, known as ***Helleborus* × *ericsmithii***, which combines the characteristics of all three parents with long-lasting flowers that appear in midwinter. It makes an excellent container plant.

OTHER FLOWERING PERENNIALS

There are a few other perennials that flower in the winter. The evergreen ***Iris unguicularis***🏆, with scented flowers in mauve and yellow, is a great plant for those who like to bring the garden indoors, as when picked the buds soon unfurl. The floppy standard petals and broader falls are rich mauve fading to white at the base, feathered and streaked with purple, and finished off with a deep yellow central stripe. The first buds will appear from early winter onwards, to be followed

Iris unguicularis 'Mary Barnard'

by a continuous succession of delicately scented flowers nestling down among the leaves. (See Good Companions, page 78.)

In the wild this iris grows around the southern Mediterranean, particularly in Algeria, where the forms with the largest flowers occur. It likes a sheltered spot, the base of a warm wall being ideal, where it can be left undisturbed. The secret is not to cosset the plant, improve the soil or add fertilizer; it does best in poor, stony, slightly alkaline soil and likes a summer baking, and the older the clump the better it flowers. A mature clump will be densely crowded with leaves, up to 60cm (2ft) high and the same across. Unfortunately the narrow strap-shaped leaves do become very tatty, spoiling the effect of the flowers. The best thing to do is to cut the foliage right back as soon as the flowers have finished and clear away all the debris at the base of the plant; new leaves will soon appear.

The forms to grow, apart from the species itself, are ***Iris unguicularis* 'Mary Barnard'**🏆, which has smaller flowers, freely produced, with narrow, rich purple petals strongly feathered down the centre; ***Iris unguicularis* 'Walter Butt'**, with paler lavender-grey flowers; and the white-flowered form ***Iris unguicularis* 'Alba'**. The closely related ***Iris lazica***🏆

Iris unguicularis 'Walter Butt'

Viola odorata

has broader leaves and flowers on shorter, 10cm (4in) stems; it prefers a damp soil in dappled shade.

Sweet violets, ***Viola odorata***, were traditionally sold in the streets in late

Viola odorata luxuriates in the shelter provided by an old tree stump. The fragrant flowers appear from late winter into early spring.

winter in wonderfully scented bunches. They were grown in cold frames to protect their delicate petals but it is quite possible to grow them outside in a sheltered spot, perhaps at the base of a hedge, on the shady side, where the perfume will catch you by surprise one late winter's day. They need a bit of cosseting, being planted in enriched soil, but if happy they will settle in and offer a few flowers for picking through the winter. The flowers of the species are deep purple; they are white in ***Viola odorata* 'Alba'** and pink in the highly scented **Rosea Group**; **'King of Violets'** is enchanting, with deep violet double flowers.

It is probably not a good idea to recommend the various forms of the wood violet, ***Viola riviniana***: its rosette-forming clumps produce many lateral stems that take root with the greatest of ease. However, in a wild part of the garden these often produce a few little violet flowers in late winter, which are a delight to pick and mix in with snowdrops and early primroses. There are pink and white forms as well. The purple-leaved violet, formerly *Viola labradorica* but more correctly ***Viola riviniana* Purpurea Group**, can be used as ground cover even in deep shade, but

GOOD COMPANIONS

The pretty strawberry-pink flowers and light green leaves of *Pulmonaria rubra*🏆 (1) are shown to advantage against the shining burgundy foliage of *Pittosporum tenuifolium* 'Tom Thumb'🏆 (2).

Although neither *Iris unguicularis*🏆 (3) nor *Nerine bowdenii*🏆 (4) has good foliage, both have delightful, delicate flowers: the nerine sugar-pink in autumn, the iris lavender-blue in winter.

although pretty it is almost impossible to eradicate once planted.

Although the main group of **pulmonarias** are spring-flowering plants, ***Pulmonaria rubra***♕ does start to flower in winter. This is a vigorous weed-smothering perennial coming into growth early in the year, with large unspotted leaves in a light mint-green and clusters of little coral-red trumpet flowers (see Good Companions, left). There are several named cultivars, all close to the species, with ***Pulmonaria rubra* 'Barfield Ruby'** said to have the largest flowers and ***Pulmonaria rubra* var. *albocorollata*** with cream flowers. These pulmonarias are ideal for planting beneath deciduous shrubs, where they will cover the ground and keep weeds at bay.

Primula vulgaris* subsp. *sibthorpii♕, from the eastern Mediterranean, is a variant of our common primrose. Its pale lilac flowers always appear in late winter and make a fine companion for snowdrops. Also flowering very early is the pure white ***Primula* 'Gigha'**, which looks delightful tucked beneath shrubs.

Hacquetia epipactis♕ is a little winter-flowering woodlander that pops up unexpectedly early in the year. It looks like a little green daisy, with a frill of bright shining green bracts surrounding a boss of minute, golden yellow florets. Once the plant has been pollinated the three-lobed, serrated leaves expand into neat hummocks, about 23cm (9in) high, and remain for the rest of the summer. In the variegated form ***Hacquetia epipactis* 'Thor'** the leaves are subtly edged with cream and the flowers are a much paler primrose-yellow. These hacquetias like a shady, humus-rich soil; they can be divided in the autumn when they are beginning to make new growth, but they need to be well labelled as the buds will not begin to show until midwinter. Coming from the central European alps, they are perfectly hardy and can cope with very cold weather, just waiting until conditions are suitable to spring into growth.

Pulmonaria rubra

Primula vulgaris subsp. *sibthorpii*

Pulmonaria rubra var. *albocorollata*

WINTER IN SUMMER

Hacquetia epipactis♕ appears flowers first, the leaves developing later on and persisting long after the blooms have faded. In the variety *Hacquetia epipactis* 'Thor' each leaf lobe is edged in creamy white.

Flowering trees and shrubs

Many winter-flowering shrubs have inconspicuous flowers but they waft a powerful scent (these are described in The Sensory Garden on pages 136–43), while others rely on their visual impact alone. These include both deciduous plants, whose seemingly fragile flowers hang daintily from bare stems, and evergreens, their handsome foliage providing a glossy backcloth for the flowers. The dwarf heathers bring welcome early colour, and there are also a few trees that are brave enough to produce their flowers in winter.

The almond-scented flowers of *Prunus mume* 'Beni-chidori' open on the bare branches in late winter. Beneath the tree the shining green arrowhead leaves of *Arum italicum* subsp. *italicum* 'Marmoratum' are marked with a tracery of silver veins.

Prunus × subhirtella 'Autumnalis'

WINTER-FLOWERING TREES

The majority of our winter-flowering trees are forms of *Prunus,* the ornamental cherry. Some, from Japan and China, were first described in the 18th century by the Swedish botanist Carl Thunberg, but only in the last 100 years have most become widely available in the West.

Prunus × subhirtella **'Autumnalis'**♈, the autumn cherry, forms a slight, open-branched tree with a spreading canopy; even when clothed with small green leaves it does not cast a lot of shade. This is a good choice for a small town garden, where the cherry blossom can be enjoyed at close quarters. It has pendulous semi-double flowers, delicate pink in bud, opening creamy white, and blooms in bursts over the winter, culminating in early spring. For rose-pink blossom, choose ***Prunus* × *subhirtella* 'Autumnalis Rosea'**♀.

Prunus mume, known as the Japanese apricot, is a small tree with scented flowers smelling of almonds. Although the flowers are usually borne in spring, in mild years they can start appearing in midwinter. The best one to grow is ***Prunus mume* 'Beni-chidori'** (which translates as 'thousand red birds'): it has wonderfully fragrant, little cup-shaped, semi-double flowers in a clear rose-pink, with a starburst of prominent cream stamens, freely borne all along the bare branches. The summer foliage is rather dull so the tree is best planted in a sunny spot at the back of a border, or trained against a wall with something more interesting in front, such as a silver buddleia, which can be hard pruned before the cherry comes into flower; tuck in a variegated arum at the base for attractive foliage from winter through to summer. ***Prunus mume* 'Pendula'** forms a small weeping tree with pale pink flowers, and ***Prunus mume* 'Alboplena'** has semi-double white flowers.

Prunus incisa, the Fuji cherry, grows slowly into a light, airy shrub or small tree with a flutter of delicate, pink-budded white flowers along its bare branches in spring. The early-flowering cultivar ***Prunus incisa* 'Praecox'**♀ has white flowers, as does the smaller, semi-double ***Prunus incisa* 'Mikinori'**. One of the most popular is ***Prunus incisa* 'Kojo-no-mai'**, a very slow-growing, angular shrub with zigzag branches and pale pink flowers – exquisite when underplanted with small blue spring bulbs. It also colours well in autumn and is even more enchanting when you know that its name means 'dance in the ancient castle' in Japanese. ***Prunus incisa* 'February Pink'** is a delicate shrubby tree with cut-edged leaves, which turn orange-red in autumn; in mild periods from late winter onwards the bare branches are studded with pale pink flowers. These cherries are not long-lived trees, suffering from a disfiguring blight, but they are still worth planting for their winter blossom.

Prunus incisa 'Praecox'

Arbutus × *andrachnoides*

Arbutus* × *andrachnoides♀ and its parent ***Arbutus unedo***♀, the strawberry tree, are small, hardy, evergreen trees that have attractive peeling bark and young branches in cinnamon-red (see page 146). They produce creamy-white flowers in late autumn and winter while the previous year's fruit still remains on the branches. The fruit resembles a strawberry, but not in taste. Although belonging to the Ericaceae family, both species will tolerate alkaline soil and so are useful for those gardening on chalk.

The cornelian cherry, ***Cornus mas,*** is a small tree or rounded shrub with

Cornus mas

EARLY FLOWERS UNDER TREES

The dappled shade under winter-flowering trees and shrubs is the perfect setting for early bulbs. *Chionodoxa luciliae*♀ **(1)**, glory-of-the-snow, is one of the first, with starry, pale sapphire flowers on stems 10cm (4in) high. *Puschkinia scilloides*, with paler blue-white flowers, is of similar stature and likes the same conditions. Both make lovely planting partners for winter aconites, with buttercup-yellow flowers surrounded with green ruffs **(2)**. Pulmonarias **(3)** and *Anemone blanda* **(4)**, snowdrops **(5)** and primroses **(6)** add to the picture.

Acacia dealbata

deciduous leaves and in late winter it is adorned with masses of small bright yellow flowers, creating an airy yellow cloud (see Good Companions, page 84). They are followed by cherry-like red fruit, edible but with a sharp flavour, used in the past for making preserves. ***Cornus mas*** **'Variegata'**♀ has striking white-margined foliage, and ***Cornus mas*** **'Aurea'** comes with a powdering of gold in the leaves.

Acacia dealbata♀, the florist's mimosa, is a lovely fast-growing evergreen tree that can be grown against a high sunny wall. It has feathery fern-like foliage and tiny bead-like buds in early winter that open to sprays of fluffy, sweetly scented flowers in mid- to late winter. In a favoured location it will grow to 10m (33ft) and is a magnificent sight when in full bloom, especially against a clear blue sky.

Jasminum nudiflorum

WINTER-FLOWERING SHRUBS

Winter jasmine, ***Jasminum nudiflorum***♀, is one of the earliest shrubs to flower, the cheerful bright yellow tubular flowers lighting up dull days in early winter. The green, leafless, arching stems are first covered with tightly furled buds and then the first splash of yellow appears, with more flowers opening spasmodically over the next couple of months. This jasmine is a native of western China and was introduced as a garden plant by Robert Fortune in 1844; he sent specimens back to Britain using the newly developed Wardian case, a closed glass container that greatly reduced the losses in shipping plants home. The Latin name aptly describes the habit of the plant, *nudiflorum* meaning naked flowers, coming on bare stems before the leaves. It quickly gained popularity for its winter colour, and it is still used in winter gardens today, valued for its toughness and longevity. Treated as a climber it can reach 5m (16ft) but, lacking any means of self-support, it needs the old stems tied in to a wall or fence and then the new stems can cascade down in a waterfall of yellow. If tightly controlled it can be trained beneath a window or allowed to

PLANTING PARTNERS FOR WINTER JASMINE

The somewhat untidy and straggly habit of *Jasminum nudiflorum*♀ can be disguised by clever planting. On a wall grow it with the large-leaved *Hedera colchica* 'Sulphur Heart'♀ (**1**): the bright yellow jasmine highlights the gold variegation in the ivy, which also provides colour and interest when the jasmine is not in flower. The best flowers are at the tips of the shoots, but the bare base of the jasmine can be hidden by an evergreen shrub such as *Mahonia aquifolium* 'Apollo'♀ (**2**): the mahonia flowers open as those of the jasmine begin to fade.

HEATHERS

The dwarf shrubby heathers offer both winter flowers and coloured foliage and are accommodating plants, usually doing best on a light sandy soil. All those described here are happy in acid conditions and tolerate alkaline soils (the summer-flowering heath, *Calluna vulgaris*, needs an acid soil – see page 115). The numerous cultivars of *Erica carnea* come in a range of colours, but one of the best is *Erica carnea* 'Myretoun Ruby'🏆 (1), which produces an abundance of rich ruby-red flowers; others are 'Winter Beauty', with pink flowers, and the earlier-flowering 'December Red'. Many varieties have attractive winter foliage, such as *Erica carnea* 'Ann Sparkes'🏆, which has yellow foliage becoming tipped with bronze-red and pinkish-purple flowers, *Erica carnea* 'Springwood White'🏆 (2), the finest white cultivar, and the old cultivar *Erica carnea* 'Vivellii'🏆, with bronze-green leaves setting off the rosy-pink flowers to perfection. These varieties need little pruning, merely cutting back to promote compact growth and prevent spreading, with a light trim to remove the spent flowers in spring. They associate well with cyclamen beneath pines.

The Irish heath, *Erica erigena,* is not quite as hardy, but the form *Erica erigena* 'Golden Lady'🏆 has white flowers on bushy, upright growth that is bright yellow all year round, and *Erica erigena* 'Irish Dusk'🏆 has dark green leaves and salmon-pink flowers.

There is a hybrid between these two species, *Erica* × *darleyensis*, which is a little taller than *Erica carnea*. *Erica* × *darleyensis* 'Arthur Johnson'🏆 (3) has long dense sprays of deep mauve-pink flowers, useful for cutting. *Erica* × *darleyensis* 'White Perfection'🏆 (4) is a good choice, a vigorous form with green leaves and white flowers, and *Erica* × *darleyensis* 'Kramer's Rote'🏆 has truly red flowers borne for a long time.

The taller tree heaths, *Erica arborea* (5), form medium to large shrubs that produce fragrant creamy-white flowers towards the end of winter. *Erica arborea* 'Albert's Gold'🏆 is most often planted and it will eventually reach 2m (6ft); its upright branches bear bright golden yellow foliage in winter. The cultivar *Erica arborea* 'Estrella Gold'🏆 is more greenish yellow and slower in growth.

PLANTING WITH HEATHERS

Although traditionally partnered with conifers, winter-flowering heathers are much more versatile plants that have many uses in the garden. Varieties with deep pinkish-red flowers, such as *Erica carnea* 'Myretoun Ruby'🏆, make a rich combination with purple heucheras and the winter foliage tints of some bergenias. They can be planted with grasses such as *Carex comans* to form a textured carpet beneath the white and salmon stems of birches. In containers they look good throughout winter and partner early *Primula* Wanda Group successfully. If kept trimmed low after flowering they are useful to disguise the fading foliage of early bulbs such as crocuses and dwarf narcissus.

Garrya elliptica 'James Roof'

grow up and over a porch. It can also be grown as a free-standing arching bush or be allowed to flow down a bank or over a low wall. It needs to be thoroughly pruned after it has finished flowering in early spring so that plenty of new flowering shoots will develop. A good combination is to grow the jasmine on a picket fence with an evergreen mound of *Viburnum davidii*🏆 in front. (See also page 124.)

Garrya elliptica graces many a sheltered shady wall. Its dark, upright form and sombre leaves are the perfect background for the spectacular display of long silver-grey catkins, which festoon the branches in late winter. The cultivar **'James Roof'**🏆 produces the longest and showiest tassels and is the best one to grow. (See also page 125.)

In a shady position the dwarf evergreen ***Ribes laurifolium*** is a useful plant: in late winter the leathery leaves and greenish-white flowers are a pleasing combination (see page 110) and a subtle companion for early-flowering bulbs.

The old winter stalwart ***Viburnum tinus***, also known as laurustinus, has been a standby of shrubberies for hundreds of years. It was introduced from the Mediterranean in the 16th century and was loved by the Victorians as a hardy evergreen, sharing a neglected corner with rhododendrons, common laurel (*Prunus laurocerasus*🏆) and spotted laurel (*Aucuba*). The evergreen leaves are a glossy dark green and the domed heads of lacy white flowers open from pink buds from late autumn onwards, pausing in midwinter and departing with a flourish at the end of winter, with shiny blue-black fruit to follow. This is an easy shrub, not fussy about soil, and left to itself can grow up to 3.5m (12ft). Although it tolerates deep shade, it flowers better in a sunny position. It benefits from having some of the old growth pruned out at the base and the tips of the branches cut back to encourage a more compact shape; this is best done in spring as soon as the flowers have faded. The colour of the buds is important since at any one time the plant can have half of its flowers just going over, with more buds coming along. By far the best form to grow is ***Viburnum tinus* 'Gwenllian'**🏆, which was introduced in the 1980s. A more compact plant with neat green leaves on red stalks, it has red-stemmed flowers that are bright pink in bud, opening white flushed with pink,

Viburnum tinus 'Gwenllian'

Viburnum tinus

GOOD COMPANIONS

Gold-variegated *Euonymus japonicus* 'Chollipo'🏆 (1) looks striking with the frothy yellow flowers of *Cornus mas* (2) in winter and provides interest in summer when the cornus flowers are over.

The grey-green red-tipped foliage of *Hebe* 'Red Edge'🏆 (3) combines well with the dark green leaves, pink buds and white flowers of *Viburnum tinus* 'Gwenllian'🏆 (4). The hebe has white flowers in summer.

OTHER WINTER-FLOWERING SHRUBS *Chimonanthus praecox* • *Daphne bholua* • *Edgeworthia chrysantha* •

with red on the backs of the petals. It also flowers in succession and often some of the indigo-blue berries are still left on the plant in the winter. If you sniff very hard, you can detect a faint, sweet scent. (See Good Companions, below left.)

Viburnum tinus **'Eve Price'**🏆 is another good compact form, as is ***Viburnum tinus*** **'Purpureum'**, with white flowers and purple young foliage. The variegated form, ***Viburnum tinus*** **'Variegatum'**, has pink leaf-stalks and creamy-white flowers, but is more tender and is best grown against a wall for protection.

The evergreen **pieris** really blossom in the spring, with bunches of pendulous, bell-shaped white flowers, but some have red buds that develop in early winter and show up well against the glossy green leaves. The very hardy ***Pieris japonica*** **'Christmas Cheer'** has pink and white flowers that often appear during winter. All need a lime-free soil and associate well with heathers. (See also pages 114–15.)

Rhododendron 'Praecox'

Pieris japonica 'Christmas Cheer'

Salix gracilistyla 'Melanostachys'

Salix udensis 'Sekka'

There are some delightful willows (***Salix***) that produce eye-catching catkins in late winter. The name catkin comes from the Tudor word for a kitten and the colloquial name for the British native willow is 'pussy willow', presumably for its resemblance to a cat's soft fur. The catkins are wonderfully soft to touch and come in various colours. Those of ***Salix gracilistyla*** **'Melanostachys'** are black, while ***Salix udensis*** **'Sekka'** has chestnut-brown catkins and curiously flattened stems, which are popular with flower arrangers. In many cases, the catkins appear on coloured stems before the leaves and these too are very good for cutting: ***Salix acutifolia*** **'Blue Streak'**🏆, for example, has dark purple stems with a blue bloom; ***Salix daphnoides*** has violet-purple stems overlaid with white, and ***Salix daphnoides*** **'Aglaia'** has plain red stems.

On acid soils some **rhododendrons** start to flower in late winter, their delicate blooms opening in milder spells of weather. They need shelter, ideally under light tree cover, as frost can damage the flowers. ***Rhododendron*** **'Praecox'**🏆 (meaning early) is one of the first, producing wide funnel-shaped, lilac-purple flowers on a small compact shrub. (See pages 113–14.)

Hamamelis × *intermedia* 'Pallida' • *Lonicera* × *purpusii* • *Sarcococca hookeriana* var. *digyna* • *Skimmia japonica* 'Rubella' •

WINTER FOLIAGE

Foliage colour in winter can be surprisingly rich and varied. True, the palette is generally quieter than that of summer, with all its bright hues, but there are many muted tones from which to choose, ranging from silver and blue to gold, cinnamon and fawn. Rich sumptuous reds and purples can also be found, as can bright emeralds and lime greens, as well as the vibrant sunshine yellow of some variegated plants. Judiciously mixing coloured leaves with the plain greens will make a lively and interesting picture to please the eye throughout the winter.

RIGHT: *Arum italicum* subsp. *italicum* 'Marmoratum' growing through *Tellima grandiflora* Rubra Group.

Green

Shrubs and trees that retain their foliage in winter come into their own as their deciduous neighbours shed their leaves. In every shade of green from palest peridot to darkest emerald, evergreens exist in every layer of the planting, from tall dark pines down through shining broad-leaved shrubs to delicate ferns and perennials nestling among the leaf litter.

Ilex × *altaclerensis* 'Camelliifolia'

Prunus laurocerasus 'Otto Luyken'

Prunus laurocerasus ETNA ('Anbri')

Choisya ternata

BIG-LEAVED EVERGREENS

Broad-leaved evergreens in plain glossy green are a mainstay of the winter garden: their foliage provides essential structure and the setting for delicate flowers and other colourful characters of the season.

The dark green hollies (***Ilex***) are fine examples. They excel in the garden just as they do in the natural landscape. ***Ilex* × *altaclerensis* 'Camelliifolia'**♀ grows into a fine pyramidal shrub with dark stems and large, spineless, shining dark green leaves. ***Ilex aquifolium* 'Ferox'**, the hedgehog holly, offers a totally different dark green texture with its bristly leaves carrying spines that stick out in every direction. It is a lower, slower-growing holly making a broad shrub 1.5m (5ft) in height. (See also pages 44–45.)

The cherry laurels, cultivars of ***Prunus laurocerasus***♀, have large, shining evergreen leaves that are particularly reflective when planted in sunny situations; most are dark holly green in shade. ***Prunus laurocerasus* 'Rotundifolia'** has shorter, broader leaves than the species and lends itself to shaping into a rounded bush. The selection ***Prunus laurocerasus* ETNA ('Anbri')**, a seedling of 'Rotundifolia', grows into a dense upright bush with bronze young leaves that mature to dark glossy green. It has proved to be extremely hardy. ***Prunus laurocerasus* 'Otto Luyken'**♀ can be clipped twice a year to keep it as a mound 1m (3ft) high; if it does grow too big, it can be cut to the ground and will soon regrow.

Choisya ternata♀, the Mexican orange blossom, has brighter, emerald-green foliage and grows into a medium-sized shrub, around 1.5m (5ft) in height and spread. By midwinter it usually produces white buds that promise fragrant white spring blooms to come. It is related to both rue and lemon, and its foliage is oily and aromatic. *Choisya ternata* grows in sun or shade. (See also page 139.)

For hot, sunny situations many varieties of **cistus** have wonderful evergreen foliage that looks good throughout the winter. ***Cistus* × *cyprius* var. *ellipticus* 'Elma'**♀ is particularly fine, with sticky leaves of dark glossy green. Growing up to 1m (3ft) tall and wide, it has saucer-shaped, soft white flowers in summer.

Bolder, darker and glossier, ***Fatsia japonica***♀ is invaluable against a shady wall. Its massive dark green leaves look dramatic and exotic and make a good contrast with smaller, variegated evergreen leaves such as those of euonymus and elaeagnus. The closely related ivies (***Hedera***) also offer many interesting green-leaved cultivars that bring emerald and dark green shades to walls and fences and form green carpets under trees and shrubs (see pages 46–47).

The foliage of ***Griselinia littoralis***♀ can provide a totally different shade of green in the garden. Rounded apple-

OTHER GOOD PLAIN GREEN EVERGREEN SHRUBS *Berberis darwinii* • *Ceanothus thyrsiflorus* 'Skylark' •

green leaves are carried on ochre-coloured stems to form a large, upright shrub that can be used to break up blocks of dark evergreens; it also mixes happily with golden-variegated ones. It is good in coastal gardens and is superb as a hedge.

Where space allows, the Chinese privet, ***Ligustrum lucidum***♕, is a magnificent large shrub or small tree with big glossy, pointed leaves and a broad conical shape. The large sprays of white flowers that appear in autumn often linger into early winter. It is a good alternative to ***Magnolia grandiflora***, which also has handsome glossy green leaves, usually with russet undersides (see page 111).

Fatsia japonica

Griselinia littoralis

Ligustrum lucidum

SMALL-LEAVED EVERGREENS

Box, ***Buxus sempervirens***♕, forms the basic green framework in many gardens, where it is used for hedging and edging. At its best it is dark, shining green; however, winter wet often causes bronzing of the foliage as nutrients are washed from the soil. In severe weather the top of low hedges of ***Buxus sempervirens* 'Suffruticosa'**♕ can be caught by frost and the leaves become white and parchment-like. (See also page 45.)

Some **hebes** bring some of the freshest evergreen leaves for the winter garden and they are solid, compact plants that can be used as specimens; they may even replace box in providing structural shape, but not as hedging because they do not respond as well to clipping. In general the hardiest ones seem to have small leaves and small, usually white summer flowers. ***Hebe* 'Emerald Gem'**♕ is a dwarf bun-shaped shrub up to 30cm (12in) tall, ideal for planting in gravel or a pot. The tiny glossy leaves are emerald green. ***Hebe rakaiensis***♕ has bright pea-green leaves, and forms a compact mound, ideal for ground cover in full sun. For topiary without clipping choose ***Hebe topiaria***♕, which grows into a grey-green dome up to 1m (3ft) high. ***Hebe macrantha***♕ is a dwarf mounded shrub with bright green, leathery leaves and relatively large pure white flowers.

For the darkest small evergreen leaves, ***Osmanthus* × *burkwoodii***♕ and ***Osmanthus delavayi***♕ are unbeatable. The former is upright, bushy and dense, with fairly shiny, small and pointed dark leaves. It grows to 1.5m (5ft) and responds well to trimming and shaping. In spring, small white fragrant flowers appear in the leaf axils. *Osmanthus delavayi* has a more spreading habit and is usually shorter in height, with tan-coloured branches and very dark, rounded, serrated leaves. Even small shrubs are smothered in fragrant white flowers in spring.

For lighter texture, rosemary, ***Rosmarinus officinalis***, is a good choice with its upright stems of grey-green, narrow, aromatic leaves (see page 152). It can be cut and shaped and makes a pleasing rounded bush up to 1m (3ft) high.

Buxus sempervirens 'Suffruticosa'

Hebe rakaiensis

Osmanthus × burkwoodii

Cistus × hybridus • Lonicera pileata • Mahonia × media 'Charity' • Osmanthus armatus • Sarcococca confusa • Viburnum tinus •

Dark green yew and holly and emerald-green hebes provide evergreen structure throughout winter and summer in this border at White Windows in Longparish, Hampshire.

GREEN CONIFERS

The British native yew, ***Taxus baccata***♕, is the ultimate dark green structure shrub and is the backbone of the English garden throughout the year. It is clipped and shaped into hedges and topiary, and its various forms make striking specimens. ***Taxus baccata*** **'Fastigiata'**♕ has the dark raven-green foliage associated with this wonderful versatile plant. (See page 50.)

Other conifers with soft sprays of foliage or bristling needles offer a different range of shapes and textures but in a similar spectrum of every possible shade of green. The exquisite ***Chamaecyparis obtusa*** **'Nana Gracilis'**♕ has shell-like sprays of foliage that resemble the folds of green velvet in shades of darkest green and brighter emerald. It grows slowly to 1.5m (5ft) in height and is a wonderful specimen plant for the rock garden or a pot. The dwarf pines have a spikier appearance: ***Pinus mugo*** **'Mops'**♕ makes a dense, dark green, bristling bush slowly reaching 1m (3ft).

Taxus baccata

Chamaecyparis obtusa 'Nana Gracilis'

EVERGREENS AT GROUND LEVEL

Dark green helianthemums and *Iberis sempervirens*♕ form low mats of neat leaves in sunny places in the garden. *Thymus serpyllum* varieties spread as dark evergreen carpets in gravel or over the edge of paving. For shade, *Asarum europaeum* has round, shining leaves just above ground level (see page 56) and *Helleborus foetidus*♕ has dark green leaves and lime-green flowers (see page 123). The variant of the wood spurge *Euphorbia amygdaloides* var. *robbiae*♕ (above) has evergreen rosettes of dark green, shiny leaves and is a very useful spreader for dry shade. It grows well with the ivy *Hedera helix* 'Manda's Crested'♕, which has waved leaves that may bronze a little in winter.

EARLY BULB FOLIAGE

Some bulb foliage appears very early in the year and with warmer winters *Ipheion uniflorum* is starting into growth in midwinter, producing fresh green grass-like clumps of leaves well before the usually pale blue flowers. The short, broad, emerald-green leaves of the late-flowering snowdrop *Galanthus ikariae* (above) bring a welcome touch of bright green at this time of year, perfectly setting off the pearly-white flowers.

FERNS

Evergreen ferns are an excellent choice for providing green in winter, their bold or feathery fronds adding different textures to the garden. The soft shield fern, *Polystichum setiferum* (**1** and **2**), forms a large clump with impressive arching feathery fronds in a soft mid-green. There are many selected forms with even more feathery, frilly fronds, some with fringed tips and others with overlapping segments so that the midrib curves round in a spiral effect. The leaves persist throughout the winter and only need to be removed when the new fronds begin to unfurl in spring. The names of these ferns can be difficult, but plants chosen from any of the many groups such as Acutilobum, Cristatum, Divisilobum or Plumosodivisilobum (**4**) will provide graceful plants for shady corners. Soft shield ferns remain in good shape all winter, will tolerate drought to a certain degree and can cope with poor soils.

A fern that often appears uninvited is *Asplenium scolopendrium* (**3**), the hart's tongue fern, which produces loose shuttlecocks of strap shaped fronds with crinkled edges. They are a bright vibrant green with a reflective surface so that they sparkle when they catch the sunlight. Grown at the back of a border, where the long narrow leaf blades can be seen when herbaceous plants have died down, it can reach as much as 60cm (2ft) high; however, it is most often found growing from cracks in shady walls, and here it is much smaller. In some cultivars the fronds are even more crimped and fringed at the tips, such as *Asplenium scolopendrium* Cristatum Group and 'Kaye's Lacerated'.

Polypodium cambricum, the southern polypody, grows in a slightly different manner from the rest of the fern family. The feathery new fronds do not appear until late summer and then they remain a fresh lettuce-green all winter. These ferns are very tolerant of dry situations and can be grown right up to the trunks of small trees, associating well with spring-flowering *Lathyrus vernus*, a dwarf vetch with fresh green, pea-like foliage and clusters of vibrant violet and cerise flowers, and the autumn-flowering *Saxifraga fortunei*, with strawberry-like rosettes of bristly, dark green silver-veined leaves. *Polypodium vulgare* (**5**), the common polypody, is a very hardy species that forms a thick evergreen carpet. It is ideal as ground cover in dry areas under trees and on banks, and grows well on chalk soils.

All these ferns will grow in either alkaline or acid soils but the Japanese shield fern, *Dryopteris erythrosora* (**6**), will thrive only in acid conditions. This is a beautiful fern for winter: grown at the foot of a cold wall, it is seemingly able to produce an unending supply of orange-brown new fronds, fading through glossy green to pale lime and arching outwards as they age. Those gardening on alkaline soil should grow it in a container using lime-free compost.

White, cream and silver variegated

Plants with variegated foliage are a valuable source of colour in any garden. They help to break up the mass of green in summer and can carry the colour of a planting scheme throughout the year even when the majority of flowers have faded. Plants with variegation in white, cream or silver pick up the light in winter and can illuminate areas that are in shadow at this time of year; fortunately there are plenty of evergreen shrubs in these colours that grow well in such a situation.

Pittosporum 'Garnettii'

The white-splashed leaves of *Euonymus fortunei* 'Emerald Gaiety' make a fresh, bright partner for snowdrops.

VARIEGATED SHRUBS

***Prunus laurocerasus* 'Castlewellan'**, a variegated cherry laurel, is a large, bushy, rounded shrub with leathery leaves heavily marbled with white. It makes a good planting partner for other evergreen shrubs with bolder white and cream variegation and it can be shaped and controlled like any of the other laurels (see page 88).

Prunus laurocerasus 'Castlewellan'

Pittosporum tenuifolium 'Irene Paterson'

Some of the striking New Zealand **pittosporums** form large columnar plants, their small shining leaves crimped at the edges. The popular ***Pittosporum* 'Garnettii'**🏆 has grey-green leaves with cream margins, speckled with red dots in cold weather, while ***Pittosporum tenuifolium* 'Silver Queen'**🏆 has smaller leaves with more cream in the patterning. ***Pittosporum tenuifolium* 'Irene Paterson'**🏆 is a pleasing selection with new leaves in pale green suffused creamy white, with black stems; it slowly grows into a medium-sized bush. Pittosporums produce tiny chocolate-coloured flowers in spring, with a lovely sweet fragrance.

The light and airy ***Osmanthus heterophyllus* 'Variegatus'**🏆 is another pleasantly variegated evergreen, with pointed leaves in two tones of green splashed with cream. It is very slow-

MORE SHRUBS WITH WHITE, CREAM AND SILVER VARIEGATED FOLIAGE *Fatsia japonica* 'Variegata' •

growing, eventually reaching 2m (6ft), and in autumn tiny cream flowers give off a strong, sweet scent. (See page 49.)

Ilex aquifolium **'Ferox Argentea'**🏆, the silver hedgehog holly, is a rather loose shrub with dark green leaves outlined in ivory; these are almost completely covered in spines, on the leaf blade as well as on the edge. In a mixed planting it has a silvery, lightening effect. ***Ilex aquifolium*** **'Handsworth New Silver'**🏆 is a much more upright, pyramid-shaped tree with silver-variegated foliage and pronounced purple stems. The narrow leaves are mottled grey and edged with cream. This is a female holly and should produce berries if a male plant is grown nearby.

Ilex aquifolium 'Ferox Argentea'

Hedera canariensis 'Gloire de Marengo'

There are plenty of silver- and white-variegated **ivies**, all of which will gently illuminate shady places, either as climbers or as ground cover. ***Hedera helix*** **'Glacier'**🏆 is widely planted, its leaves a varying mixture of grey and green with lighter silver-grey patches and a thin cream rim. It will grow to 3m (10ft) or more on a wall or fence or up a tree. ***Hedera canariensis*** **'Gloire de Marengo'**🏆 is often grown as a houseplant but makes a light variegated subject for a sheltered spot outside, where it will grow to 3m (10ft). The large leaves are deep green in the centre marbled with silver-grey and edged with white.

Like ivy, the sprawling evergreen ***Euonymus fortunei*** cultivars are useful as ground cover or against a wall, and there are a number of patterned forms. ***Euonymus fortunei*** **'Emerald Gaiety'**🏆 has bright white margins to the small, dark green leaves, which flush with pink in cold weather. (See also pages 116–17.)

For acid soils there are **rhododendrons** and **pieris** with variegated leaves. ***Rhododendron ponticum*** **'Silver Edge'**, slower-growing than the plain green form and usually reaching about 2m (6ft), has distinctive white margins to the leaves. The soft lavender-blue flowers appear in late spring and early summer. ***Pieris japonica*** **'White Rim'**🏆 is a slow-growing shrub reaching 1.5m (5ft), with narrow leaves edged in creamy white and pink-flushed when young. ***Pieris japonica*** **'Little Heath'**🏆 has much smaller leaves with a cream rim, and is more compact in habit (see also page 114).

Pieris japonica 'Little Heath'

Euphorbia characias SILVER SWAN ('Wilcott')

In recent years several **euphorbias** with silver-variegated foliage have been introduced although many have proved to be less hardy than the plain-leaved forms. ***Euphorbia characias*** SILVER SWAN (**'Wilcott'**) is one of the best selections, forming a bushy shrub with grey-green linear leaves outlined in ivory. It grows to around 1m (3ft) in height and makes a wonderful focal point wherever in the garden it is planted.

VARIEGATION AND GROWTH

The green colour in a leaf comes from the pigment chlorophyll, which is essential for plant growth. Because there is less chlorophyll in variegated foliage than in plain green leaves, the variegated plant will often grow less strongly and be smaller than the plain green form – a fact that can be useful in a small garden.

REVERTING TO GREEN

Variegated plants rarely occur in the wild; most originate and are cosseted in gardens, and therefore demand a little more care than the species with plain green leaves. When growing variegated plants, particularly shrubs and trees, watch for any shoots producing plain green leaves and cut them out, right at the base. They need to be removed quickly, as they are more vigorous and will soon overwhelm the stems with patterned leaves.

Hebe × *franciscana* 'Variegata' • *Griselinia littoralis* 'Dixon's Cream' • *Rhamnus alaternus* 'Argenteovariegata' •

USING VARIEGATED FOLIAGE PLANTS

Evergreen plants with variegated leaves – with the base green broken up by stripes, splashes or spots of white, silver, cream or yellow – can add lightness and definition to beds and borders in all seasons and bring interest to a dark corner.

There are several points to consider when choosing a variegated plant. A bold stripe down the centre of a leaf or a neat outline to the edge is more eye-catching than irregular splashes or an overall mottling or speckling. *Prunus laurocerasus* 'Castlewellan' (**1**), for example, has white-marbled leaves and is better used as a backdrop to a border than viewed close to, while hollies with leaves neatly margined in white, cream or gold are good both close up and at a distance; *Ilex aquifolium* 'Argentea Marginata'🏆 (**2**) is particularly fine, with each prickle outlined in white.

Variegated plants need to be carefully placed, and too many used too close together can be discordant; plenty of plain green is always needed as a mixer. If using several variegated plants choose ones with contrasting leaf shapes. The broad silver-spotted leaf of one of the pulmonarias (**3**) looks well next to the smaller oval leaf of a silver and green euonymus such as *Euonymus fortunei* 'Emerald Gaiety'🏆 (**4**), whereas the euonymus next to a variegated periwinkle, which has a very similar leaf, would not be so successful.

Ideally, place variegated plants among plain green companions or plants that are all gold or all silver. Plants with white or cream markings, such as *Euphorbia characias* SILVER SWAN ('Wilcott') (**5**), go best with those with silver foliage, such as *Astelia chathamica*🏆 (**6**); those in gold or bright yellow are better with gold foliage. A bamboo and ivy, both with sharp yellow in their leaves, would be excellent companions (see page 98), but pairing the bamboo with a silver-patterned ivy could be less effective.

Variegated plants will still be making an impact in the garden in other seasons, so companions should be chosen with this in mind. A cream-variegated euonymus is a perfect partner for *Physocarpus opulifolius* 'Dart's Gold'🏆 (**7**), with butter-yellow leaves in summer, as in winter the rich chestnut stems of the physocarpus stand out against the green and cream leaves of the euonymus.

VARIEGATED PERENNIALS

Perennials with overwintering leaves are very useful growing beneath bare shrubs. The little deadnettle ***Lamium maculatum* 'White Nancy'**🏆, if cut hard back in early autumn, will produce fresh leaves for the winter in pale silver with a narrow green edge. The white flowers are a perfect complement in summer. Although vigorous in some situations, this deadnettle is easily controlled by trimming twice a year.

Lamium maculatum 'White Nancy'

The shade-loving ***Chiastophyllum oppositifolium* 'Jim's Pride'** forms low-growing rosettes of succulent, rounded leaves in cream and green, some all cream, flushing pink in winter, with arching sprays of tiny yellow flowers in summer. Winter aconites look charming nestling among the foliage.

Chiastophyllum oppositifolium 'Jim's Pride'

***Carex conica* 'Snowline'** is a neat little evergreen sedge with narrow dark green leaves edged in white. ***Carex morrowii* 'Variegata'** is slightly larger, with pale cream edges to the leaves. For dramatic sword-shaped foliage in deep green banded with cream, choose ***Iris foetidissima* 'Variegata'**🏆, whose spiky clumps look attractive all through the year (see page 61). The lilyturf ***Liriope spicata* 'Gin-ryu'** (sometimes sold as 'Silver Dragon') has narrower grassy leaves with a silver stripe; it forms a dense mat, providing excellent ground cover in sun or light shade, with the benefit of dense spikes of pale violet flowers in late summer.

Carex conica 'Snowline'

Galactites tomentosa

Arum italicum subsp. *italicum* 'Marmoratum'

The thistle-like ***Eryngium variifolium*** has overwintering basal leaves with conspicuous white veins, making a striking foil for the spiky, straw-coloured seedheads. ***Galactites tomentosa*** is a hardy annual thistle, with deeply cut green leaves margined in white and either purple or white flowers in summer; if a few plants are allowed to set seed, little plantlets will have formed by the winter, startling in their combinations of silver and green.

The marbled foliage of ***Arum italicum* subsp. *italicum* 'Marmoratum'**🏆 is an invaluable component of the winter garden, setting off the first snowdrops and still looking fresh for the spring narcissus. This low-growing tuberous perennial is a back to front plant, looking its best in winter, flowering unobtrusively in spring and dying back in summer, then producing spikes of bright orange berries in the autumn before the new foliage appears (see page 27). The mid- to dark green leaves are shaped like arrowheads, with the midribs and veins brightly etched in ivory. The birds relish the berries, helping to move the plant around the garden; it usually pops up unexpectedly beneath deciduous shrubs, where it is seen in winter and goes unnoticed in summer. In their first year the seedling leaves are generally green, only acquiring their smart markings when older.

MULTICOLOURED FOLIAGE

An increasing number of plants with multicoloured leaves are now available from nurseries and garden centres. They can certainly help to enliven a winter planting, but the mixture of colours does not appeal to all gardeners. They work well in the garden as long as the constituent colours of the foliage are taken into consideration when choosing neighbouring plants.

SHRUBS

***Leucothoe fontanesiana* 'Rainbow'** Low, arching shrub with leaves variegated with cream, yellow and pink. (See page 114.)

***Nandina domestica* 'Fire Power'**🏆 Small shrub, with broad ferny leaves, yellowish green in summer, taking on shades of orange and red in winter.

***Osmanthus heterophyllus* 'Goshiki'** (1) Compact shrub with small holly-like yellow-mottled green leaves, bronze-tinted when young.

Pseudowintera colorata (2) Small, mounded shrub with aromatic, leathery leaves in pale yellow, flushed pink and blotched dark crimson-purple.

PERSISTENT PERENNIALS

***Ajuga reptans* 'Multicolor'** Overwintering leaves in a mixture of rose and purple, splashed with cream.

***Bergenia cordifolia* 'Tubby Andrews'** (3) Large paddle-shaped leaves in cream and green in summer, taking on shades of cherry, pink and apricot in winter.

***Heuchera* CRÈME BRÛLÉ ('Tnheu041')** (4) A recent introduction with gently waved leaves in old gold flushed with copper, pink and purple on the reverse. (See page 59.)

***Tiarella* 'Skid's Variegated'** Summer leaves in apple green suffused with cream, turning coral pink in winter. (See page 60.)

Evergold

Whether in the form of flowers or foliage, the colour yellow always attracts the eye, so careful positioning in the garden is important. Gold foliage can dominate the quieter tones of winter; however, like golden spring and summer flowers, it can lift the planting and bring sunshine to flowerbeds and borders. Use it to lighten heavy evergreens and to accent gold-variegated shrubs such as aucubas and elaeagnus.

Choisya ternata SUNDANCE ('Lich')

Ilex crenata 'Golden Gem'

Choisya GOLDFINGERS ('Limo')1

Hebe ochracea 'James Stirling'

Lonicera nitida 'Baggesen's Gold'

Erica carnea 'Foxhollow'

Choisya ternata SUNDANCE (**'Lich'**)♀, a variant of the popular Mexican orange blossom (see page 88), forms a solid, rounded, medium-sized shrub, with glossy, brilliant yellow foliage that turns a paler yellow in cold weather. It is useful for providing an eye-catching patch of colour, and tends to look best combined with contrasting foliage in plain green or with gold-variegated shrubs. The foliage is a softer, more lime yellow when planted in shade. ***Choisya*** GOLDFINGERS (**'Limo'**) is less vigorous, with finer foliage, and has an altogether softer appearance.

The tiny leaves of the shrubby honeysuckle ***Lonicera nitida* 'Baggesen's Gold'**♀ are golden yellow, turning a lighter yellow-green in winter, although the colour is not so bright if grown in poor soil or in a shady site. This is a graceful shrub, 1.5m (5ft) or more high, with gently arching branches densely covered with foliage. Left to itself it will form a rather shaggy mound, but it responds very well to clipping. It makes a soft backdrop for summer flowers, especially those in pure yellow, and is excellent in containers.

Conversely, some golden shrubs intensify in colour as winter advances. ***Ilex crenata* 'Golden Gem'**♀, for example, a small, low-growing holly with neat golden yellow leaves, deepens to a burnished old gold. Some of the hardy whipcord **hebes**, with conifer-like foliage on neat, star-shaped, low bushes, become a burnished gold in winter; ***Hebe ochracea* 'James Stirling'**♀ is the one most often planted, with rich ochre-gold foliage and tiny white flowers along the branches in summer. Many of the **heathers** have brilliant golden foliage, intensifying in colour in the winter: The foliage of ***Erica carnea* 'Foxhollow'**♀ becomes rich yellow tinged red in winter, setting off the white flowers; ***Erica carnea* 'Ann Sparkes'**♀ is slow-growing and spreading, with gold foliage and purple-pink flowers.

Some conifers also take on richer tones, such as the useful yellowish-green *Thuja*

MORE SHRUBS WITH GOLD FOLIAGE *Abies nordmanniana* 'Golden Spreader' • *Erica vagans* 'Valerie Proudley' •

occidentalis 'Rheingold'🏆, which turns a rich copper-gold in winter (see page 51). It has a loose, informal habit that mixes happily with broad-leaved shrubs. ***Platycladus orientalis*** **'Aurea Nana'**🏆 (formerly classified as *Thuja*) is golden green taking on hints of bronze. This is a conifer with a particularly good shape: its flattened foliage sprays are displayed almost vertically, slowly forming a perfect cone up to about 2m (6ft) high.

The golden-leaved bay, ***Laurus nobilis*** **'Aurea'**🏆, is a large evergreen structure shrub forming a broad cone up to 5m (16ft) or more in height. The foliage is greenish yellow throughout summer, turning gold in winter.

Platycladus orientalis 'Aurea Nana'

Laurus nobilis 'Aurea'

BIG YELLOW CONIFERS

Big yellow conifers can look out of place in a country garden, but provide exceptionally bold structure when used as focal points. Good ones include *Chamaecyparis lawsoniana* 'Stardust'🏆, growing into a large feathery cone up to 10m (33ft) in height, with soft sprays of golden yellow foliage, bronze at the tips. *Chamaecyparis lawsoniana* 'Lutea'🏆 is an old cultivar of broad, columnar habit with large flattened sprays of golden yellow leaves. *Chamaecyparis lawsoniana* 'Winston Churchill' (above) is denser and broader than these, with golden yellow foliage throughout the year.

For an overall lighter effect, in terms of structure, try ***Pittosporum tenuifolium*** **'Warnham Gold'**🏆. This forms a tall, narrow columnar shrub with dense, softly wavy greenish-yellow foliage that gradually becomes more golden through the year; in winter when caught by the sun it is a glistening buttery yellow. It is an excellent shrub that deserves wider planting and suits coastal gardens.

Hedera helix **'Amberwaves'** is a most attractive ivy, with dainty shallowly lobed leaves in pure pale gold. It looks lovely trailing around the base of deciduous shrubs or mingling among evergreens. The ivy's winter colouring fades to soft chartreuse in the summer.

The little perennial golden feverfew, ***Tanacetum parthenium*** **'Aureum'**, has ferny leaves in bright lime yellow, at their sharpest in spring but holding their colour well even in winter. The sprays of orange-eyed, white daisy flowers are freely borne all summer long, leaving seedlings behind to pop up in unexpected places. The plant settles happily for full sun or even dry shade, where it shows up well, but it is less satisfactory in heavy, damp soil.

More splashes of winter gold come from grasses and sedges. ***Acorus gramineus*** **'Ogon'** has angled fans of fine-pointed leaves, which on close inspection are seen to be striped green and yellow; the overall effect is a sharp acidic citrus. The wintergreen woodrush, ***Luzula sylvatica*** **'Aurea'**, has mounds of wide-bladed leaves turning an iridescent gold in winter. In early spring the coffee-coloured spikelets show up well against the foliage. It likes a rich, moist, leafy soil in shade.

Pittosporum tenuifolium 'Warnham Gold'

Luzula sylvatica 'Aurea'

Escallonia laevis 'Gold Brian' • *Ilex aquifolium* 'Flavescens' • *Ligustrum ovalifolium* 'Aureum' •

Gold variegated

Gold-variegated plants can be used to add bright highlights to plantings of darker evergreens. They make striking combinations with evergreens with golden foliage and are cheerful partners for early yellow narcissus and crocuses. They are an essential constituent of any yellow planting scheme, helping to maintain the colour theme throughout the seasons with or without the presence of flowers.

Euonymus fortunei 'Emerald 'n' Gold'

Vinca minor 'Illumination'

The gold and green striped leaves of *Pleioblastus auricomus* are sharp and vibrant against the soft green and buttercream foliage of the variegated ivy *Hedera helix* 'Goldchild' growing beneath.

There are innumerable gold-variegated hollies (***Ilex***) with varying patterns of green and gold on the leaves. Probably the two best and most widely planted are ***Ilex* × *altaclerensis* 'Golden King'**🏆 and ***Ilex* × *altaclerensis* 'Lawsoniana'**🏆. 'Golden King' has large, almost spineless leaves of dark green with broad gold margins; in 'Lawsoniana' the variegation is reversed, with the gold appearing in the centre of the leaf. Both produce red-brown berries in profusion and form large conical shrubs. They make wonderful specimen plants and provide excellent structure at the back of a large border. (See also pages 44–45.)

Ilex × *altaclerensis* 'Golden King'

The eye-catching ***Euonymus fortunei*** BLONDY (**'Interbolwi'**), with a primrose-yellow centre to each green leaf, is one of several gold-variegated euonymus, which form good ground-covering shrubs for sun or shade. In some the colouring can be a little strident, and these need positioning with care. The very popular ***Euonymus fortunei* 'Emerald 'n' Gold'**🏆 changes colour through the seasons: in summer the leaf margins are bright yellow-gold, becoming a darker more antique gold in winter, tinged with pink when planted in sun. (See also page 117.)

There are also many gold- and yellow-variegated **vincas**. The most striking of these is ***Vinca minor* 'Illumination'**, with dark green leaves with bright golden yellow at the centre; the pale blue flowers are of secondary importance. It is an eye-catching plant and works well in containers and contemporary planting schemes with trimmed box shapes. ***Vinca minor* 'Aureovariegata'** has golden-edged green leaves and similar blue flowers.

Some of the variegated **elaeagnus** grow rapidly into large shrubs with rather harsh yellow patterning. The widely grown ***Elaeagnus pungens* 'Maculata'**, with bright gold blotched leaves, is suited

GRASSES, BAMBOOS AND PERENNIALS WITH GOLD-VARIEGATED FOLIAGE *Carex oshimensis* 'Evergold' •

Elaeagnus pungens 'Maculata'

Elaeagnus pungens 'Frederici'

Aucuba japonica 'Pepperpot'

Hedera colchica 'Dentata Variegata'

to large gardens and also responds well to pruning if this is done lightly from an early age. It is prone to reversion so plain green shoots should be removed when seen; some caution is needed here as the variegation does not show on the youngest leaves. The more compact ***Elaeagnus pungens* 'Frederici'**, with narrow leaves in creamy yellow with a green border, is a better choice for smaller gardens. It forms a bushy shrub up to 1.5m (5ft) in height. Elaeagnus are very tolerant shrubs and cope well in exposed and coastal situations.

Opinions differ as to the merits of the spotted **aucubas**. For some, the irregular gold blotches and spots on the shining green leaves are decidedly unappealing, but few would deny that ***Aucuba japonica* 'Golden King'**🏆 is a striking sight in semi-shade, with its glossy emerald leaves liberally spotted with gold. ***Aucuba japonica* 'Marmorata'** is an improvement on 'Golden King'; it has equally strong markings but will take more sun. ***Aucuba japonica* 'Pepperpot'** is a compact variety with smaller leaves finely sprinkled with gold. If male and female plants are grown, bright red berries will result on some cultivars. (See also page 116.)

There are plenty of gold-variegated **ivies** from which to choose. The much-planted ***Hedera helix* 'Goldheart'** (officially named 'Oro di Bogliasco'🏆), with three-lobed leaves with a large yellow splash in the middle, is a vigorous plant soon covering a wall. ***Hedera helix* 'Goldchild'**🏆 is a weak variety as a pot plant but strong-growing and successful when grown against a wall. The green-marbled leaves are edged with gold. Selections of ***Hedera colchica***🏆, with large, generally unlobed, heart-shaped leaves are ideal for clothing a wall, although they may need tying in to wires, wall nails or a framework to get them started because their suckering rootlets are not as effective as those of the common ivy. ***Hedera colchica* 'Dentata Variegata'**🏆 has dark green leaves attractively outlined with creamy gold, and ***Hedera colchica* 'Sulphur Heart'**🏆 (formerly 'Paddy's Pride') has dramatic leaves in light and dark green with bold splashes of yellow in the centre.

OTHER GOOD GOLD-VARIEGATED ELAEAGNUS

The variegated elaeagnus are popular shrubs and have given rise to some of the most widely planted variegated evergreens.

***Elaeagnus × ebbingei* 'Limelight'** (above) Large, spreading shrub with upward-sweeping branches carrying grey-green leaves with soft yellow centres. In early winter insignificant creamy-white flowers emit a delicious fragrance that will permeate the garden.

***Elaeagnus × ebbingei* 'Gilt Edge'**🏆 Altogether brighter with dark green leaves with a broad gold margin. It is slower-growing but very striking.

***Elaeagnus pungens* 'Goldrim'**🏆 Deep glossy green leaves edged with yellow.

PLANTING PARTNERS

The gold-variegated leaves of *Hedera colchica* 'Sulphur Heart'🏆 contrast in form with those of *Lonicera nitida* 'Baggesen's Gold'🏆 but complement the colour.

Cortaderia selloana 'Aureolineata' • *Liriope muscari* 'Variegata' • *Saxifraga* 'Aureopunctata' • *Tolmiea menziesii* 'Taff's Gold' •

Brown, tan and orange

Shades of brown predominate in the winter landscape, in the bare earth, in the naked stems and twigs of deciduous trees and shrubs and in the fallen leaves that still litter the ground. In the case of beech and hornbeam the dried, coppery-brown leaves cling to the branches, catching and filtering the winter sun. Perennials and grasses with chestnut and bronze foliage are often suffused with pink and orange or deepened to a rich chocolate brown, and the low light in winter accentuates these tones.

The fluted and curled leaves of hornbeam, *Carpinus betulus*, turn to copper-brown and cling on to the twigs throughout winter.

The tough **epimediums** (see page 55) provide shades of bronze and chestnut with their overwintering foliage, useful for brightening up a woodland corner; however, these leafy perennials are too good to be relegated to the woodland floor as they fare equally well in a border, where their new leaves, often green edged with bronze, make pleasant leafy mounds in the summer. ***Epimedium davidii*** turns an unusual shade of ochre and chestnut in winter, with rounded, scalloped, prickle-edged leaves and yellow flowers in spring, while ***Epimedium*** × ***rubrum***♀ fades to a rich mahogany brown.

The foliage of some of the acid-loving, summer-flowering heathers turns orange and gold in winter: ***Calluna vulgaris* 'Joy Vanstone'**♀ has golden foliage becoming a rich orange, **'Wickwar Flame'**♀ turns a fiery russet red, and **'Ariadne'** is bright gold in summer turning to shades of red and copper in winter.

Several sedges have leaves the colour of milk chocolate and one of the very best is ***Carex comans* 'Bronze'**, a froth of bronzy brown with undertones of pink, most

Calluna vulgaris 'Ariadne'

striking in the winter. It can be allowed to fountain out in a tangle, but can also be hard pruned twice a year and it makes a better winter foliage plant in this form.

Equally attractive are the smaller, brownish-red ***Carex comans* 'Taranaki'** and the larger, finely leaved ***Carex buchananii***♀. All these sedges are perfect in dappled shade, but they will also be happy in sun if the site is not too dry. There are two more sedges with

PLANTING PARTNERS

Grow several *Carex comans* 'Bronze' with *Tellima grandiflora* Rubra Group (**1**) at the base of the Manchurian cherry, *Prunus maackii* 'Amber Beauty'♀ (**2**), or the river birch, *Betula nigra*: the carex will echo the colour of the tree bark.

OTHER BROWN WINTER FOLIAGE PLANTS *Anemanthele lessoniana* • *Corokia* × *virgata* 'Frosted Chocolate' •

leaves in shades of olive green and tan. Of these ***Carex dipsacea*** has the richest orange-tan leaves, which take on even brighter tones in winter, especially in sun; ***Carex testacea*** has narrower leaves in the same combinations. They are evergreen and grow into mounds of arching foliage, making good partners for bergenias in the light shade of birch trees.

The best coloured sedge to grow is ***Uncinia rubra***, with its tussocks of stiff blades in tones of crimson and mahogany, the perfect foil for tall snowdrops.

***Hakonechloa macra* 'Aureola'**♕ has the appearance of a soft dwarf bamboo. It forms a clump of delicate stems, with arching leaves of bright yellow streaked green. In winter these remain on the plant, becoming dry and parchment-like and assuming shades of rich corn.

Persicaria affinis (*Polygonum affine*) is not always considered as a winter foliage plant but this autumn-flowering knotweed forms mats of creeping stems from which arise narrow dark green leaves. After the first frosty nights these turn several shades of deep reddish brown and remain as a warm carpet all winter. The narrow bottlebrush flowers too will fade to a pale russet. The cultivars to consider are ***Persicaria affinis* 'Darjeeling Red'**♕, with rosy-red flowers, and **'Donald Lowndes'**♕ in deep pink.

Libertia peregrinans has evergreen fans of iris-like leaves, whose central veins are stained orange and yellow, giving the whole plant a sunset glow, especially in winter (see page 118). Clusters of white flowers are borne among the leaves in early summer. It will grow in any well-drained soil, in either sun or light shade. It looks stunning planted beneath a witch hazel, interspersed with snowdrops.

The celandine ***Ranunculus ficaria* 'Brazen Hussy'** (see page 68) has chocolate-brown leaves that appear just in time to complement the flowers of the winter cyclamen; as the cyclamen blooms are fading, the celandine begins to open its own bright yellow flowers.

Carex buchananii

Uncinia rubra

Hakonechloa macra 'Aureola'

PRUNING CAREX

To prune fine-leaved carex, gather up the leaves and stems in one hand and, using a pair of scissors (secateurs do not work), cut off the top third, including the long flowering stems. This will leave the plant arching out gracefully, but not trailing along the ground. It may be necessary to do this twice a year, at the beginning and end of the summer.

GROUND-COVER CONIFERS WITH BROWN WINTER FOLIAGE

Many conifers show bronzing of the foliage in winter. *Microbiota decussata*♕ (right) resembles a prostrate juniper with soft sprays of pale green foliage that becomes bronze in the colder months. It forms an attractive carpet beneath blue conifers, particularly the blue spruce *Picea pungens* 'Globosa'♕. The Canadian juniper, *Juniperus communis* var. *depressa*, forms spreading mats of brown-green foliage that turns bronze in winter. *Juniperus communis* 'Repanda'♕ also shows good winter colour when planted in full sun.

Heuchera 'Caramel' • *Phormium* 'Surfer Bronze' • *Podocarpus nivalis* 'Bronze' • *Thuja occidentalis* 'Ericoides' •

Red, purple and black

Rich red, purple and black hues add warmth and depth to the winter planting. Cold weather and low winter light intensify the shades of those plants that normally have foliage of these colours. In other shrubs and perennials the colours develop as the days grow shorter and colder. These leaf tones are ones to use to enhance delicate winter flowers or the colourful bare stems of deciduous shrubs.

Pittosporum tenuifolium 'Tom Thumb'

Mahonia aquifolium 'Atropurpurea'

Tellima grandiflora Rubra Group

Leucothoe LOVITA ('Zebonard')

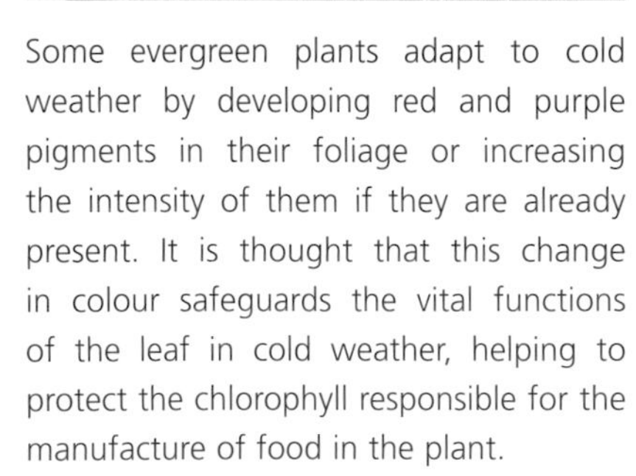

Some evergreen plants adapt to cold weather by developing red and purple pigments in their foliage or increasing the intensity of them if they are already present. It is thought that this change in colour safeguards the vital functions of the leaf in cold weather, helping to protect the chlorophyll responsible for the manufacture of food in the plant.

One of the most striking shrubs to do this is ***Pittosporum tenuifolium* 'Tom Thumb'**🏆, with foliage that becomes richly purple in winter, the sun reflecting off the glossy, crinkled leaves. This grows into a small, dense, rounded bush with the new young leaves opening green but changing to reddish purple as the season progresses. It makes an excellent companion for red-stemmed dogwoods or pale cream hellebore flowers. It is also an excellent subject for a pot.

The handsome glossy green foliage of some **mahonias** takes on gorgeous reddish tints in winter, especially if grown in sun. The leaves of ***Mahonia aquifolium*** and its cultivars will turn a glowing shade of reddish purple, while those of ***Mahonia japonica***🏆 develop rich tints of orange and red, particularly strong on poor, well-drained soil. (See also pages 48 and 142.)

Some conifers turn purple in winter. Selections of ***Cryptomeria japonica* Elegans Group** are bushy tall shrubs or small trees with soft feathery foliage of sea green that turns red-bronze in winter (see page 50). The slower-growing, smaller ***Cryptomeria japonica* 'Elegans Nana'** forms a billowing bush of soft green turning purple in winter. ***Cryptomeria japonica* 'Vilmoriniana'**🏆 is a dwarf form with tightly packed foliage turning red-purple in winter. It is suitable for a rock garden.

The large-leaved **bergenias** (see page 54) give a perfect demonstration of colour change, going from bright green to dramatic tones of mahogany, claret and rhubarb red when exposed to sun in winter; and the heuchera relative ***Tellima grandiflora* Rubra Group** turns a softer coral red (see page 60).

The acid-loving evergreen ***Leucothoe*** LOVITA **('Zebonard')** is a compact shrub with arching stems carrying red-tinted leaves that intensify to reddish purple in winter. The less arching ***Leucothoe*** SCARLETTA **('Zeblid')** has leaves that become scarlet at the tips, and ***Leucothoe axillaris* 'Curly Red'** has broad leaves that curl back on themselves, a rich red in winter, on a neat bushy plant. All are excellent for containers if potted in lime-free compost. (See also page 114.)

Some **heucheras** have crinkled leaves that are almost black, shot through with silver, in cultivars such as ***Heuchera*** EBONY AND IVORY **('E and I')**, **'Obsidian'** and **'Beauty Colour'**. ***Heuchera* 'Stormy Seas'** is vigorous with purple leaves that

PURPLE-LEAVED EVERGREENS *Lophomyrtus* × *ralphii* 'Kathryn' • *Loropetalum chinense* f. *rubrum* 'Fire Dance' •

Hebe 'Red Edge'

Ophiopogon planiscapus 'Nigrescens'

Cornus alba 'Kesselringii'

remain fresh in all weather, and the robust ***Heuchera* 'Chocolate Ruffles'** has ruffled leaves in dark chocolate-purple with burgundy undersides. (See pages 58–59.)

***Hebe* 'Red Edge'**♀, a hybrid of *Hebe albicans*♀ (see page 105), is a small neat rounded bush with pleasing grey leaves outlined in maroon. In winter the maroon colour intensifies, particularly at the tips of the branches, giving an overall impression of a plum-coloured bush. ***Hebe* 'Caledonia'**♀ has attractive plum-tinted foliage on a dwarf bush.

Many **euphorbias** take on rich tones of purple in winter and the newly introduced ***Euphorbia* BLACKBIRD ('Nothowlee')** is proving a winner, growing no more than 45cm (18in) high with narrow leaves in shades of grape purple and green, crimson at the tips (see page 130).

Selections of ***Phormium tenax* Purpureum Group**♀ make bold winter specimens providing a useful change of form, with spiky clumps of broad arching leaves, reaching 1.5m (5ft) or more high, in shades of purple and bronze. Grow them in a sunny border to rise up through low, mounded plants. At a lower level the dwarf forms, such as ***Phormium tenax* 'Nanum Purpureum'**, with red-purple leaves, ***Phormium* 'Tom Thumb'**, with bronze foliage, and the almost black ***Phormium* 'Platt's Black'**, are more suitable. (See also page 52.)

***Ajuga reptans* 'Atropurpurea'** is a ground-covering carpeter with overwintering rosettes of burnished purple leaves and dense spikes of blue flowers in spring. It is best planted where it gets some sun to enhance the colour, but it will perish if it gets too dry. ***Ajuga pyramidalis* 'Metallica Crispa'** is a miniature form with wonderful scrunched and curled foliage offering texture in winter and reflecting the sun. It combines well with the evergreen *Vinca minor* 'Illumination', with its bright gold-centred leaves.

The grass-like ***Ophiopogon planiscapus* 'Nigrescens'**♀ is one of the very few plants with almost black leaves, but it does need to be carefully placed. It hardly shows up at all when grown in bare brown soil, but it looks striking against a gravel path or partnered with silver plants. It also makes a delightful companion for bright red or pale pink *Cyclamen coum*♀ (see page 56) and it is a good choice for a pot.

The deciduous maidenhair fern ***Adiantum venustum***♀ sheds its leaves to reveal wiry black stems, which form an unusual backdrop for a small green-flowered hellebore and a clump of snowdrops. ***Cornus alba* 'Kesselringii'**♀ has purple-black stems and works well partnering other dogwoods with stems in reds and yellows. It is also stunning when planted against the white-barked Himalayan birch, *Betula utilis* var. *jacquemontii*♀ (see page 145).

Ajuga reptans 'Atropurpurea' forms a shining wine-red carpet beneath snowdrops alongside a shady path.

Pittosporum tenuifolium 'Purpureum' • *Rhododendron* 'Joanna' • *Rhododendron* PJM Group 'Peter John Mazitt' •

Eversilver and blue

Silver foliage plants are often associated with dry, sunny conditions and warm weather; however, they show up all the more in the darker days of winter, contrasting with dark evergreens and bare branches. Those with felty grey leaves are particularly beautiful when etched with frost. Plants with blue foliage are striking partners for gold- and white-variegated evergreens; blue conifers make eye-catching specimens; and blue grasses and euphorbias mix well with purple heucheras and bergenias.

Cynara cardunculus

Helichrysum italicum 'Korma'

SILVER FOLIAGE PLANTS

Lavenders are a natural choice for the winter garden and a well-pruned bush of narrow silver leaves shines out in midwinter. All thrive in a sunny, well-drained site, tending to be short-lived in heavier soils. Some of the best lavenders are the ***Lavandula* × *chaytorae*** hybrids, which inherit a very silvery-white leaf from *Lavandula lanata* and winter hardiness from *Lavandula angustifolia*. ***Lavandula* × *chaytorae* 'Sawyers'** is an excellent silver form with large, pointed, deep purple flower spikes, growing to about 50cm (20in) high (see page 118), and ***Lavandula* × *chaytorae* 'Richard Gray'** is a little smaller; both have the bonus of scented foliage. Prune lavenders immediately after flowering in late summer, cutting back to where new shoots are visible; the new growth will develop into neat hummocks, remaining in good shape all winter.

Another old favourite for a hot, dry spot is the cotton lavender, ***Santolina chamaecyparissus***, a small dense bush with the most delightful tiny, filigree, silver-grey leaves (see page 118). It should be pruned after flowering in midsummer, shaping and cutting back to new growth each year, to keep the plant dense and bushy. The summer flowers come in pale cream or bright yellow. There is also a green-leaved cultivar, ***Santolina rosmarinifolia* subsp. *rosmarinifolia***, which contrasts well with the silver forms.

Helichrysum italicum, the curry plant, has long, narrow silver leaves and a more open habit than santolina and lavender; this is an advantage in damp winter weather as the foliage does not become waterlogged. It forms a low shrub up to 60cm (2ft) tall, with similar spread. ***Helichrysum italicum* 'Korma'** is brighter with excellent silver foliage. Helichrysum should be cut back after flowering in late

PROTECTING SILVER-LEAVED PLANTS FROM WINTER WET

Most downy, silver-leaved shrubs and perennials come originally from the Mediterranean area and have adapted to grow happily in that hot, dry climate. They do well enough in colder regions, but tend to languish if winters are too wet. To improve their chances of survival, give them a really free-draining position, adding plenty of grit to heavy soil, and surround them with a layer of grit or gravel to stop rain splashing on to the leaves. Clearing away leaf litter in late autumn will also help prevent silver-leaved plants becoming too soggy.

summer; this will result in a flush of new silver growth that will stay looking good throughout the winter.

The grey-leaved ***Brachyglottis* 'Sunshine'**🏆 (more familiar under its former name *Senecio*) comes from Dunedin in the far south of New Zealand and has proved a very hardy evergreen. In time it can grow into a large shrub, but it can be pruned to shape and, if desired, all the bright acid-yellow daisy flowers, borne in summer, can be removed. Close inspection reveals that the grey leaves are outlined in white and the underside covered in a white felt. ***Brachyglottis* 'Walberton's Silver Dormouse'** is a newer cultivar that is much smaller, and could be grown in a container.

Brachyglottis 'Sunshine'

PLANTING PARTNERS

Silver foliage plants such as *Brachyglottis* 'Sunshine'🏆 add bright interest to the winter border. Here the shape and size of the brachyglottis leaves are a pleasing contrast to the small bright green leaves and compact habit of the hebe behind.

Atriplex halimus, the tree purslane, is a good eversilver shrub with shining, silky, oval leaves (see page 118). Although vigorous, reaching 3m (10ft) or more, it can be pruned to keep it within limits, and it is happy in poor soil in full sun. Enjoying similar conditions, ***Phlomis italica***, with pale lilac-pink summer flowers, retains its furry grey leaves in a mild winter. It forms a suckering shrub up to 1m (3ft) high with upright stems. ***Phlomis fruticosa***🏆 (Jerusalem sage) is hardier, with grey-green leaves on upright branches (see page 118). Yellow sage-like flowers appear in the leaf axils in summer.

Brachyglottis 'Walberton's Silver Dormouse'

Phlomis italica

Some of the smaller **hebes** have grey or blue-grey foliage. ***Hebe pinguifolia* 'Pagei'**🏆 is one of the most popular. It makes a mounded mat of small, neat silver leaves, and tiny white flowers appear in clusters at the end of the branchlets in summer. The taller ***Hebe albicans***🏆 and ***Hebe recurva* 'Boughton Silver'**🏆 form rounded mounds of similar silver foliage, up to 60cm (2ft) high. The leaves are larger and pointed, the flowers white and produced in summer. In larger beds they are effective planted in groups with upright clumps of miscanthus or the colourful stems of *Cornus alba* 'Sibirica'🏆.

Some **helianthemums** have pretty silver-grey foliage: their tiny leaves are carried on slender, lax stems forming low mats or mounds, and they thrive in any sunny position, looking especially good at the edge of a path or beside steps. ***Helianthemum* 'The Bride'**🏆 is a fine variety with white flowers in early summer. Plants should be trimmed after flowering to promote a flush of new growth and a bushy habit.

Hebe albicans

Helianthemum 'The Bride'

The bold silver-green leaves of ***Cynara cardunculus***🏆, the cardoon, stand up to the winter weather in milder regions. Although they reach up to 1m (3ft) long, they collapse at the end of summer when the massive branching flower stems, 2m (6ft) high, are at their peak. If the stems and leaves are all cut back to ground level as the flowers fade, new foliage will

Cyclamen hederifolium 'Silver Cloud'

Verbascum bombyciferum

emerge, upright and arching, that will stay silver throughout the winter.

Lambs' ears, ***Stachys byzantina***, also has winter-resistant leaves in a soft furry silver-grey (see page 118), and slowly forms carpets in a dry, sunny site. Woolly spikes of purple flowers appear in summer, but these soon deteriorate and are perhaps best removed. Also down at ground level there are selected forms of the autumn-flowering cyclamen, such as ***Cyclamen hederifolium* 'Silver Cloud'**, that have almost completely silver leaves, which persist well through the winter.

Young plants of the biennial ***Verbascum bombyciferum*** are equally rewarding with their winter rosettes of bright silver leaves, and in summer they are graced with tall spikes of yellow flowers. If space permits these can be left to provide seed for birds in winter; however, on light soils verbascums will seed themselves prolifically, and seedlings will need weeding out to prevent them becoming a nuisance.

SILVER IN THE SHADE

Few silver-leaved plants enjoy shade, but *Astelia chathamica*♕ is an exception. It forms a clump of upright rosettes of bright metallic silver, sword-shaped leaves, somewhat resembling the top of a pineapple. It grows well in a pot or rising out of gravel, which prevents soil splashing onto its beautiful foliage. (See also page 53.)

BLUE FOLIAGE PLANTS

Perhaps surprisingly there are quite a few blue-leaved plants for the winter garden, although there is a fine dividing line between blue and grey. All these blue foliage plants need full sun to colour well and are not suited to growing in shade.

Several of the **junipers** have grey-blue foliage. One of the bluest is ***Juniperus chinensis* 'Pyramidalis'**♕, with a good

Juniperus squamata 'Blue Carpet'

columnar shape; it is dense and slow-growing, to 2m (6ft) high. Pencil-slim ***Juniperus scopulorum* 'Blue Arrow'** and **'Skyrocket'** grow slowly to no more than 4m (13ft), so are good choices for adding height but not bulk to a small garden; both have exceedingly narrow upright growth, which is useful as a focal point. 'Blue Arrow' has steel-blue foliage of better quality than that of the duller 'Skyrocket'. Some ***Juniperus virginiana*** selections have foliage that intensifies to purplish blue in winter: ***Juniperus virginiana* 'Burkii'**, growing to 3m (10ft) high in ten years, is an excellent columnar form of dense, compact habit, with steel-blue foliage turning purple in winter.

At ground level ***Juniperus squamata* 'Blue Carpet'** has blue-grey leaves on a mat of spreading branches, and the compact ***Juniperus squamata* 'Blue Star'** forms a dwarf bun of silver-blue quartz-like foliage. For a neat, perfectly

Juniperus squamata 'Blue Star'

conical specimen for a small garden, scree bed or pot, choose ***Picea glauca*** ALBERTA BLUE (**'Haal'**), with densely packed silver-blue needles. It makes a wonderful focal point in a bed of heathers.

The so-called blue hollies, forms of ***Ilex × meserveae***, were developed by Kathleen Meserve in the 1950s on Long Island in an attempt to produce a holly that could withstand the cold winters of eastern North America. They are broad shrubs with angular shoots and glossy, dark blue-green foliage with soft spines.

MORE PLANTS WITH SILVER AND BLUE FOLIAGE *Calluna vulgaris* 'Silver Queen' • *Convolvulus cneorum* •

They look well in contemporary gardens and are good partners for silver foliage shrubs. Like *Ilex aquifolium*, one of the parents, they can be trimmed and shaped and their compact habit and smaller leaves make this easier. ***Ilex* × *meserveae*** BLUE PRINCESS, rarely more than 2m (6ft) high, has dark stems and dark foliage studded with bright scarlet berries in winter. ***Ilex* × *meserveae*** BLUE PRINCE is an excellent male form and needed as a pollinator to ensure a good crop of berries on any of the female clones.

Ilex × meserveae BLUE PRINCE

Ruta graveolens 'Jackman's Blue'

Blue rue, ***Ruta graveolens***, adds a flat metallic blue to the winter palette and is usually seen in the form of **'Jackman's Blue'.** It is a shrubby herb, reaching 60cm (2ft) in height and spread, with blue-green stems and prettily divided foliage of intense blue-grey. Mustard-coloured flowers appear in summer. It should be treated with caution as it can cause skin irritation if handled in sunlight. There is also a blue-leaved snowdrop, ***Galanthus elwesii***🏆, which has broad leaves in a rich glaucous blue, and thrives in a sunnier spot than do the green-leaved snowdrops.

The brightest blue-grey foliage, in summer and winter, comes in ***Cupressus arizonica* var. *glabra* 'Blue Ice'**🏆, a slow-growing conical tree reaching 3m (10ft) in time. It produces delightful little round cones in matching blue.

Picea pungens cultivars are among the most magnificent blue foliage plants in the winter garden, with bristling branches of silver-blue needles. ***Picea pungens* 'Koster'**🏆 is the most popular blue spruce, with a regular conical shape reaching 3m (10ft) in ten years. For smaller spaces choose ***Picea pungens* 'Globosa'**🏆, which grows to a rounded bush up to 1m (3ft) high. It makes a striking planting partner for the sword-shaped leaves of purple or yellow phormiums.

Euphorbias are fine blue foliage plants, from the shrubby forms of ***Euphorbia characias* subsp. *wulfenii***🏆 in dark grey-blue to ***Euphorbia nicaeensis*** in

Cupressus arizonica var. *glabra* 'Blue Ice'

Picea pungens 'Koster'

Euphorbia characias subsp. *wulfenii*

a brilliant almost aquamarine blue, with upright stems turning pink in winter, and the smaller-leaved, compact ***Euphorbia* 'Blue Haze'**. These all need to be grown in full sun to bring out the sharpest foliage colour. At ground level ***Euphorbia myrsinites***🏆 will snake its blue stems over paving or a low wall (see page 57).

GRASSES WITH BLUE WINTER FOLIAGE

Among the grasses some of the fescues provide good blue-grey leaves. *Festuca glauca* 'Elijah Blue' (left) forms small neat tussocks of rolled leaves and should be planted in groups in an open sunny site. Plants are not long-lived and in spring need to be trimmed back above the crowns or pulled apart and renewed. *Festuca valesiaca* 'Silbersee' (SILVER SEA) is an even brighter, almost powder blue.

Eucalyptus perriniana • *Hebe pimeloides* 'Quicksilver' • *Ozothamnus rosmarinifolius* 'Silver Jubilee' • *Teucrium fruticans* •

SITUATIONS

Many gardens offer a variety of growing conditions: some parts may be open and sunny, others shaded by trees or buildings; there may be a very wet area, or a very dry, stony site, perhaps close to the house. In some places the soil may be acid, in others it may be more neutral or even limy. There are plants that will thrive in each one of these situations; others will do only moderately well, and some will definitely perish – and matching plant to place is undoubtedly the key to a successful winter garden.

RIGHT: Winter aconites and snowdrops brighten a shady corner.

Clay soil

Clay soils are some of the most difficult to work, baking hard and cracking in the summer and remaining cold and sticky in winter, but they can be improved by adding horticultural grit to help to break up the solid lumps of clay, along with humus and strawy manure to improve the texture. Over time this will produce a fine workable tilth and, because clay soils are naturally nutritious, many plants will settle in well and need little further attention.

Ribes sanguineum WHITE ICICLE ('Ubric')

Ribes laurifolium

The low-growing evergreen currant ***Ribes laurifolium***, with drooping racemes of greenish-white flowers in late winter, flourishes in heavy clay in partial shade. A prostrate shrub, it seems happiest draping itself over a bank, and it can be covered with a not too vigorous clematis for summer flowers – a viticella hybrid such as *Clematis* 'Little Nell' is a good choice, with white flowers edged in pale pink.

Cotoneasters succeed on any soil and many, such as the graceful *Cotoneaster franchetii*, hang on to leaves and berries that cheer the winter garden.

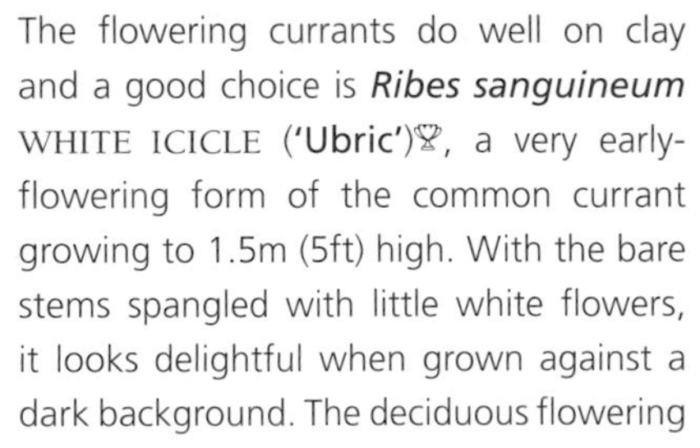

The flowering currants do well on clay and a good choice is ***Ribes sanguineum*** WHITE ICICLE (**'Ubric'**)♕, a very early-flowering form of the common currant growing to 1.5m (5ft) high. With the bare stems spangled with little white flowers, it looks delightful when grown against a dark background. The deciduous flowering currants are pruned after flowering, removing some of the older stems back to the base of the plant. This promotes vigorous straight new branches that show off the flowers to greatest advantage.

PLANTING IN HEAVY CLAY

Those gardening on very heavy clay soil should ideally plant trees and shrubs early in the autumn to give them time to establish before the winter. To avoid waterlogging, the best advice is to dig only a shallow hole – just one spade deep – but make it at least 1m (3ft) wide, and use a garden fork to make drainage holes in the bottom and sides. Rather than adding compost or manure to the planting hole, spread it as a generous layer on the soil surface.

WINTER BULBS, GRASSES AND PERENNIALS FOR CLAY *Bergenia purpurascens* • *Carex comans* •

THE NATURE OF CLAY

Clay soils are not the opposite of chalk soils or of acid soils: they can vary in pH (acidity or alkalinity) from quite acid to quite alkaline. Hardy hybrid rhododendrons will grow happily on acid clay soils, and alkaline clay soils will accommodate most of the basic evergreen shrubs that also thrive on chalk, such as holly, aucuba, cotoneaster and box (see pages 26, 44–46,116 and 124).

The proportion of fine clay particles in the soil governs how heavy and sticky it is in winter, and how difficult it is to cultivate; the more clay, the denser the soil. Digging and leaving the soil rough exposes it to frost action in cold areas. This helps to break up the clay, giving the soil a more open character. Applying garden lime has a similar effect by helping the clay particles to stick together, but this does raise the soil pH, making it more alkaline.

Leycesteria formosa🏆, pheasant berry, is often overlooked as an ornamental shrub for winter interest, but it is then, after the leaves have fallen, that the attractive sea-green stems can best be seen. A light trim to remove the twiggy tops of the upright branches, 2m (6ft) tall and arching at the tips, leaves a striking plant that will combine brilliantly with red-, yellow- and flame-stemmed dogwood (*Cornus*), which also thrive on clay, even in wet conditions. In summer, leycesteria carries drooping clusters of claret bracts and white flowers, followed by dark red berries that are particularly attractive to pheasants, hence the common name.

Leycesteria formosa

Cornus alba 'Sibirica'

The colourful willows (***Salix***) are also happy on clay, especially with a bit of moisture as well. There are plenty to choose from and many have interesting catkins (see page 85 and 120).

***Brachyglottis* 'Sunshine'**🏆 (formerly known as *Senecio* 'Sunshine') performs well on clay in sun, with pale grey leaves backed with parchment white. This is a potentially large shrub, but it can be kept hard pruned if required. (See page 105.)

The unusually shrubby ***Eupatorium ligustrinum***🏆 from Mexico has proved surprisingly hardy in recent milder winters. It forms a medium-sized rounded evergreen shrub with small, narrow, shining green leaves and lacy sprays of pale pink flowers on dark stems in the autumn. The flowers fade to a russet colour and remain on the plant over winter.

Most **magnolias** will grow on clay soils whether they are acid or slightly alkaline. ***Magnolia grandiflora***, with its large, leathery evergreen leaves, often with russet undersides, does well on clay that is not too wet in the winter, and it is often planted close to the walls of a house. Although it will eventually grow into a large tree, it can be pruned to size and increasingly is trained into a more formally shaped shrub, often at the expense of the opulent summer flowers. ***Magnolia grandiflora* 'Little Gem'** is a more dwarf variety, which has smaller leaves of darkest green with chestnut-felted undersides. It is more suitable for the smaller garden, growing slowly to 3m (10ft) or so.

Magnolia grandiflora

Magnolia stellata

Many deciduous magnolias excel in the winter garden, their pointed, soft grey furry buds catching the low light. On the bare silver branches these are every bit as attractive as the showy flowers that follow. ***Magnolia stellata***🏆 is one of the best for winter buds (see page 32), and these open to starry white flowers in early spring. With a smaller, more compact habit, normally growing to 2m (6ft) in ten years, it is the best choice for the average garden, where the ever-popular *Magnolia* × *soulangeana* often gets far too big.

Acid soil

Those who garden on acid soil are fortunate in being able to grow a wide range of shrubs to provide foliage and flowers in the winter. Rhododendrons and camellias bring in bright flowers, and gaultherias (pernettyas) provide a mass of colourful berries. Some of the pieris and many of the leucothoes have attractive leaves and, on a smaller scale, some forms of heather also have excellent winter foliage.

The late-winter flowers of *Camellia* × *williamsii* 'Saint Ewe' are remarkably resilient, even in frost or snow.

Camellia sasanqua 'Crimson King'

Camellia 'Cornish Snow'

Camellias are the natural choice of flowering shrub for all gardens on acid soil. The ***Camellia sasanqua*** varieties that flower in winter have richly perfumed flowers but are slightly tender and only suited to mild localities. ***Camellia sasanqua*** **'Narumigata'** has small cup-shaped flowers in creamy white tinged with pink; it grows into a large shrub when mature, but it can be trimmed after flowering. This one does flower outside in Britain, as does the bright red ***Camellia sasanqua*** **'Crimson King'**🏆, but others need the protection of a wall, or can be grown in large pots and brought into a conservatory or porch over winter.

Varieties of ***Camellia japonica*** are hardier and, if planted where the morning sun does not damage frosted flowers, they are a magnificent sight in early spring, although they may start blooming from midwinter onwards. ***Camellia japonica*** **'Nobilissima'** has luxuriant peony-form, waxy white flowers shaded with cream against dark green foliage. It is one of the earliest to flower. ***Camellia japonica*** **'Apple Blossom'**♡ is very appealing, with shell-pink flowers.

For smaller gardens or containers the compact and upright ***Camellia* × *williamsii* 'Jury's Yellow'**♡, with delightful anemone-centred flowers in white and pale yellow, is a good choice. This usually flowers later, at the end of winter into early spring. The lovely single-flowered ***Camellia* × *williamsii* 'Saint Ewe'**♡ is a hardy hybrid with weather-resistant rose-pink flowers that often find themselves sprinkled with snow in the gardens of southern England.

Camellia* × *vernalis has fragrant white flowers from late winter on, and the aptly named ***Camellia* × *vernalis* 'Yuletide'** has small scarlet flowers. The old hybrid ***Camellia* 'Cornish Snow'**♡ is a very worthwhile informal shrub, with masses of small white flowers in late winter.

A few evergreen **rhododendrons** flower early in the year but they do need to be planted in a sheltered spot as they dislike cold winds and too much sun in summer. The very old hybrid ***Rhododendron* 'Nobleanum Venustum'** is a big shrub bearing trusses of funnel-shaped pink flowers with darker internal markings, opening in midwinter and

Rhododendron 'Nobleanum Venustum'

continuing intermittently until spring. A few stems cut off in bud will slowly open if brought indoors. Another old hybrid called ***Rhododendron* 'Christmas Cheer'** is a compact shrub that usually blooms in late winter and early spring, with flowers pink in bud opening white; in the past it would be potted and forced into flower in time for Christmas.

The deciduous ***Rhododendron mucronulatum*** is one of the first in bloom, a slender, upright shrub with magenta flowers massed on the bare branches. If these are damaged by frost it will wait until the conditions improve, then produce new flowers in succession. ***Rhododendron mucronulatum* 'Cornell Pink'**♡ bears softer pink flowers.

There is a group of semi-evergreen rhododendrons that flower in midwinter

Rhododendron dauricum

and are a charming sight in a woodland setting. The species ***Rhododendron dauricum*** is fully hardy and easily grown and it copes well with cold weather by replacing any frosted growth. It forms a small to medium-sized, rounded shrub with an open twiggy habit, and its small trusses of flowers open from midwinter

Rhododendron mucronulatum 'Cornell Pink'

onwards in lilacs, pale pinks and even white. ***Rhododendron dauricum* 'Midwinter'**♡ has delightful tubular flowers in a clear phlox-purple, with prominent bright pink stamens. This is a rare colour in winter, and when planted in the shelter of deciduous trees with a carpet of snowdrops at its feet this

PLANTING PARTNERS FOR WINTER-FLOWERING RHODODENDRONS

As the early-flowering rhododendrons are mostly blue-pink, lilac or purple in colour, they are enhanced by being underplanted with dwarf bulbs in similar or complementary shades. *Cyclamen coum*♡ (1), with its cerise, pink or lavender flowers, is a natural partner, as are blue *Scilla siberica*♡ (2), the paler *Chionodoxa luciliae*♡ and blue forms of *Anemone blanda*♡ (3). A pulmonaria with blue or blue and pink flowers would also be a good choice. The dark foliage and lime-green flowers of *Helleborus foetidus*♡ (4) provide a sympathetic background to the rhododendrons' open habit and delicate blooms.

Rhododendron 'Olive' glows in the late afternoon sun against a backdrop of birch stems and grasses.

rhododendron is one of the finest of winter-flowering shrubs. ***Rhododendron* 'Olive'** is a lovely hybrid of *Rhododendron dauricum* with small dark green leaves and funnel-shaped crimson-purple flowers that open in profusion in late winter.

ACID-LOVING PLANTS AND ALKALINE SOIL

There is a complex relationship between soil pH (scale of acidity and alkalinity) and the availability of nutrients to certain plants. On alkaline soils certain nutrients, iron in particular, become locked up with the soil particles, and so are unavailable to plants. Plants that thrive on alkaline soil have adapted to cope with this deficiency, but in ericaceous and other acid-loving plants, the foliage begins to turn yellow, growth slows or stops and, in the worst cases, death fairly soon follows.

Although measures can be taken to change the pH of soil, such remedies never last and are not the ideal solution. Those gardening on alkaline soils should grow acid-loving plants in containers or raised beds containing ericaceous or lime-free soil or compost, and reserve the open ground for plants that prefer alkaline conditions.

Leucothoes are wonderful low-growing foliage shrubs that enjoy the

Leucothoe fontanesiana 'Rainbow'

dappled shade of trees and thrive on acid soils that are reasonably moist in summer. ***Leucothoe fontanesiana* 'Rainbow'** has arching stems and pointed oval leaves, dark green on the lower part of the branches, wonderfully marbled in cream, yellow and pink nearer the tips. The American selection **'Girard's Rainbow'** is similar, both forming mounded shrubs up to 1m (3ft) high and the same across. A number of lower-growing hybrids with neat dark green leaves that turn rich scarlet and wine in winter were originally developed by a nurseryman in Ohio. These eventually form spreading shrubs up to 60cm (2ft) high with arching shoots and make wonderful ground cover, but they need sufficient light to develop good winter leaf colour. Young plants are also very attractive planted in pots of lime-free compost to bring colour near to the house in winter. ***Leucothoe* SCARLETTA ('Zeblid')** has leaves that turn bright scarlet at the tips, and ***Leucothoe* LOVITA ('Zebonard')** turns deep crimson-purple. (See also page 102.)

Pieris are mostly grown for their showy new growth in spring, which appears in shades of peach, pink, orange and red, and for their dainty, white or pink flowers. Some form large shrubs, but most grown in gardens are compact, rounded bushes

Pieris japonica 'Little Heath'

with narrow evergreen leaves, either plain and glossy green or variegated with cream and white. The flower buds appear in early winter and hang elegantly at the tips of the stems as delicately beaded sprays, like those of lily-of-the-valley. ***Pieris japonica* 'Little Heath'**♈ has small leaves neatly edged in cream, with crimson leaf stalks. ***Pieris japonica* 'Christmas Cheer'** sometimes produces early flowers that open in winter, cream flushed with rose-pink at the tips (see page 85). ***Pieris***

OTHER WINTER SHRUBS FOR ACID SOIL *Camellia* 'Cornish Spring' • *Drimys winteri* • *Hamamelis* 'Brevipetala' •

japonica 'Katsura' is a fine, relatively new variety with dark green foliage and most attractive dark pink flower buds in winter. Like leucothoes, pieris make excellent subjects for pots.

There are some excellent berrying shrubs for acid soil, particularly the low-growing evergreens often referred to as pernettyas but now more correctly called **gaultherias**. They come from South America but are perfectly hardy and they form wiry, dense shrubs not more than 90cm (3ft) high, with a spread of 1.5m (5ft); the little, bell-like white flowers in early summer are followed by a mass of berries, which persist for most of the winter. They need to be planted in groups with male and female forms for plenty of berries, which come in a whole range of colours from white through shades of pink to cochineal, magenta and deep purple. ***Gaultheria mucronata* 'Wintertime'**♀ has the largest pure white berries, and ***Gaultheria mucronata* 'Bell's Seedling'**♀, with dark red berries, is self-fertile so does not need a male pollinator. ***Gaultheria procumbens***♀ is a good ground-cover plant with creeping stems spreading to form an evergreen rug under trees. The juicy ruby berries show up well among the leaves, which become purple-tinted with the onset of cold weather. Young plants are wonderful for winter containers and make good partners for *Skimmia japonica* 'Rubella'♀.

Gaultheria mucronata 'Bell's Seedling'

Itea virginica 'Henry's Garnet'

***Itea virginica* 'Henry's Garnet'** is a real gem for those gardening on acid soil. A loose shrub up to 1m (3ft) tall with fine tan-coloured branches, it has bright green, long narrow leaves and a mass of white bottlebrush flowers in summer. In autumn the leaves start to turn to rich crimson, staying on the plant as winter progresses and assuming a shade of deep garnet red before they finally fall in late winter.

The Japanese maples, cultivars of ***Acer palmatum***, like a well-drained, acid soil in lightly dappled shade. Although they shed their leaves in winter, they form delicate umbrella-shaped mounds of fine, twiggy branches, often with bark in tones of coral and pink. (See page 146.)

The slender, low-growing bog rosemary, ***Andromeda polifolia***, loves an acid soil and is one of the few evergreens to enjoy damp conditions in winter. The narrow grey-green leaves are carried on upright stems, giving the plant the appearance of a low bush of *Rosmarinus officinalis*. The clusters of bell-shaped pink or white flowers add colour in the spring.

While some heathers do well in either acid or alkaline soil (see page 83), the European heather or ling, ***Calluna vulgaris***, must be grown in acid conditions. These heathers flower in summer, but many varieties have good winter leaves, which intensify in colour as the temperature drops. ***Calluna vulgaris* 'Beoley Gold'**♀ has tawny yellow foliage and white flowers, while ***Calluna vulgaris* 'Robert Chapman'**♀ has soft purple flowers set against golden summer foliage, which becomes a fiery red in winter. In contrast ***Calluna vulgaris* 'Silver Fox'** has silver leaves and white flowers, and ***Calluna vulgaris* 'Wickwar Flame'**♀, with lavender flowers, has burnt-orange foliage in summer turning to a bright flame red later in the year. These summer heathers need pruning in early spring before new growth begins.

PLANTING PARTNERS FOR CALLUNAS

All the varieties of *Erica carnea* and *Erica × darleyensis* that tolerate chalk soils will be even happier on acid soils, making them natural winter-flowering partners for callunas that have interesting winter foliage. Both callunas and ericas are often associated with conifers and their colour is usually better in acid conditions. The dwarf pine *Pinus mugo* 'Winter Gold', with bright golden foliage (see page 51), makes a wonderful combination with *Calluna vulgaris* 'Robert Chapman'♀ and the dark foliage and rich ruby-red flowers of *Erica carnea* 'Myretoun Ruby'♀ (above). The attractive bristly blue foliage of *Picea pungens* 'Globosa'♀ is intensified in combination with the white flowers of *Erica carnea* 'Springwood White'♀ (see page 83) and the soft silver foliage of *Calluna vulgaris* 'Silver Fox'.

Pieris japonica 'Valley Valentine' • *Rhododendron* 'Praecox' • *Sarcococca confusa* • *Skimmia × confusa* 'Kew Green' •

Chalky soil

Those who garden on limestone or chalky soil need not despair as more plants thrive in alkaline conditions than on very acid soils – although true lime-haters such as the rhododendrons, camellias and calluna heathers cannot be grown. Most chalky soils are free-draining, which is a great advantage in winter since it means plants will rarely be waterlogged and the ground can be dug at any time. The soil also tends to warm up quickly, enabling spring growth to get off to a good start.

Daphniphyllum himalaense subsp. *macropodum*

Photinia × *fraseri* 'Red Robin'

Aucuba japonica 'Crotonifolia'

There are plenty of evergreen shrubs to consider for chalky soil. ***Daphniphyllum himalaense*** **subsp.** ***macropodum*** is grown for its whorls of pale green, rhododendron-like leaves with a glaucous reverse, drooping from red stalks. Although it grows into a small broad tree, it can be hard pruned to keep it at 2.5m (8ft).

If you want bright winter colour on a large, evergreen structural shrub then a **photinia** would be a good choice, perhaps ***Photinia*** **×** ***fraseri*** **'Red Robin'**♡, a selection from New Zealand with brilliant red new growth turning to glossy green in summer. Frequent pruning keeps the new growth coming. Alternatively, choose ***Photinia*** **'Redstart'**, with narrower leaves and a vigorous, upright habit. The new growth is scarlet, and clusters of white early-summer flowers are followed by orange-red fruit in autumn.

Other evergreens for winter interest are the spotted **aucubas**, which were overused in dank shrubberies in the past but are now being reconsidered for their cheerful glossy leaves heavily speckled with gold, as in ***Aucuba japonica*** **'Crotonifolia'**♡ or the even more striking ***Aucuba japonica*** **'Golden King'**♡. If male and female plants are grown, the females will produce bright red berries. Aucubas grow into broad, bushy shrubs 2m (6ft) or more high. They are plants for all-year interest and excellent background shrubs. The taller varieties, such as ***Aucuba japonica*** **'Variegata'**, are good for hedging and screening, particularly under trees. More compact forms make good container plants. All are tolerant plants, coping with heavy shade and atmospheric pollution. Occasionally the tips of the branches blacken in late winter and early spring; this is normally due to drought in the preceding summer rather than to frost damage. (See also page 99.)

All **euonymus** thrive on chalk. The variegated varieties of *Euonymus fortunei* with neat, oval evergreen leaves can provide colourful ground cover, be encouraged to climb on walls or fences, or be trimmed to form low, mounded shrubs. ***Euonymus fortunei*** **'Emerald Gaiety'**♡, with cream-variegated leaves (see page 92), and ***Euonymus fortunei*** **'Emerald 'n'**

IMPROVING CHALKY SOILS

Chalky soils can be improved by the addition of organic matter, which builds up the humus content of the soil, increasing its capacity to hold water and nutrients. Well-rotted manure and garden compost can be forked into the soil in spring or spread on the surface in autumn. Fresh manure should be avoided since this can supply too much nitrogen, resulting in soft growth, which may be susceptible to frost damage.

OTHER PLANTS FOR CHALKY SOIL *Buxus sempervirens* • *Clematis cirrhosa* var. *balearica* • *Cotoneaster franchetii* •

Gold'🏆, with gold and green variegation (see page 98), are particularly useful as dwarf shrubs and as low climbers at the base of shaded walls. Both flush pink in winter when grown in a sunnier position. ***Euonymus fortunei*** **'Silver Queen'** has slightly larger leaves and more conspicuous creamy-white variegation. It is slower-growing than the other varieties but is perhaps the finest when mature. The upright ***Euonymus japonicus*** **'Chollipo'**🏆 forms a dense shrub up to 2m (6ft) high, with dark green leaves that are boldly edged with gold. ***Euonymus japonicus*** **'Latifolius Albomarginatus'** has large, dark green leaves edged with white.

Daphnes are the prime choice for a winter-flowering shrub for chalky soil, relishing the free-draining conditions. The British native ***Daphne mezereum***, making a rounded bush 1m (3ft) high, bears fragrant, purplish-red flowers held tight against the bare upright branches throughout the winter months, associating well with underplantings of hellebores or snowdrops. The flowers are followed by poisonous red berries. There is also a white-flowered form, ***Daphne mezereum*** **f.** ***alba***, which has yellow fruit. These daphnes are not long-lived, seldom managing more than 10–15 years; happily, birds distribute the berries so that seedlings will appear unexpectedly and can be used as replacement plants. The shade-loving spurge laurel, ***Daphne laureola*** (see page 122), has yellow-green flowers in clusters at the tips of the branches and produces its scent in the early evening. A few twigs of any of these daphnes picked for the house produce a rich spicy fragrance that will fill a room, and the winter garden would be incomplete without their sweet scent. (See also pages 137–38.)

When it comes to perennials and bulbs for alkaline soils then all the **hellebores** are happy (see pages 72–77) and **snowdrops** and **winter aconites** will spread readily. The pinks, ***Dianthus***, bring good blue foliage, with scented flowers in the summer. The Algerian ***Iris unguicularis***🏆 likes poor, dry soil in sun (see page 77), and the dwarf bulbous irises, too, thrive in a hot, dry spot and their flowers will be much more strongly scented there. The evergreen ***Iris foetidissima***🏆 is happy in dry shade (see page 27), as are the **brunneras**, with forget-me-not blue flowers very early in spring and, in many cases, spectacularly patterned and silvered leaves later on (see page 33).

Euonymus fortunei 'Silver Queen'

Euonymus japonicus 'Chollipo'

Daphne mezereum

Dianthus

Iris unguicularis

Iris foetidissima

Hebe 'Red Edge' • *Ilex aquifolium* 'Ferox' • *Lavandula angustifolia* • *Pyracantha* 'Orange Glow' • *Rosmarinus officinalis* •

Sandy soil

Sandy soil, although rather thin and lacking in nutrients, is always easy to work in winter. It can be improved by adding organic material such as garden compost and farmyard manure, which will feed the soil as well as bulking it out and making it better able to retain moisture. However, some plants tend to make rapid growth in sandy soil and then fade away because their roots have failed to find enough sustenance at lower levels. For best results choose undemanding plants that have adjusted to growing in poor soil and dry conditions.

Atriplex halimus

Santolina chamaecyparissus

Stachys byzantina

Phlomis fruticosa

Lavandula × chaytorae 'Sawyers'

Libertia peregrinans

Many silver-leaved plants love the poor, dry conditions of sandy soil. The silky grey shrub ***Atriplex halimus*** is lovely in winter with the sun shining on its aluminium leaves. It is quite a vigorous grower, but takes readily to pruning to keep it within bounds, even twice a year. The silver-grey shrubby **phlomis** are also good candidates for sandy soil, the hardiest being ***Phlomis fruticosa***♕ (Jerusalem sage), with matt grey-green leaves but rather harsh yellow flowers in summer. The less hardy ***Phlomis italica*** has prettier, lilac-pink flowers and is best treated like a hardy fuchsia and cut back to the ground in early spring; it will soon regrow. The silver **santolinas** or cotton lavenders form attractive mounds of filigree foliage when grown in the sun, **lavender** is all the better in a hot, sandy spot, and the creeping perennial ***Stachys byzantina*** (lambs' ears) makes good ground cover here. (See pages 104–106.)

Bergenias (see pages 54–55) are adaptable perennials that can cope with dry, sandy soil, their foliage perhaps colouring better when plants are hard pressed, and some of the Mediterranean **euphorbias** offer glaucous blue foliage over winter (see page 107). Add more silver with a shrubby evergrey **helianthemum** (see page 105), cream and green in the overwintering basal growth of the thistle-like ***Eryngium variifolium,*** and fresh green on some of

OTHER PLANTS FOR SANDY SOIL *Cornus alba* 'Elegantissima' • *Erica × darleyensis* • *Helichrysum italicum* •

Sedum spectabile

the **hebes** (see page 89), and you will have a delightful garden all winter. Further colour comes with the iris-like ***Libertia peregrinans***, with fans of narrow leaves in orange and tan (see page 101).

Although not really evergreen the clump-forming **sedums** are still interesting in winter. If left uncut the flat autumn seedheads remain until well into the new year and by then there will be signs of life in the rosettes of resting buds – icy blue in ***Sedum spectabile***♀, apple green in ***Sedum spectabile* 'Iceberg'**, and dark maroon in ***Sedum* 'Purple Emperor'**. When the seedheads have finally crumbled, tidy them away and enjoy the new leafy growth. These sedums do best in a light, sandy soil, where the flowerheads will not grow too large.

Acanthus spinosus♀, the handsome perennial known as bear's breeches, forms a statuesque plant with glossy, deeply cut leaves that persist during the winter in milder areas. In severe weather the

Prunus laurocerasus 'Otto Luyken'

Rosa virginiana

foliage can be knocked back but it will emerge again the following spring. This plant must be very carefully positioned where the invasive roots cannot spread. once established, it is hard to eradicate.

Sandy soil in shade can prove difficult for many plants, but the cherry laurel, ***Prunus laurocerasus***♀, is remarkably tolerant of such challenging conditions, and a well-clipped rounded bush of ***Prunus laurocerasus* 'Otto Luyken'**♀ would provide a satisfying shape all year round. (See page 88.)

The invaluable shrub rose ***Rosa virginiana***♀ is happy in sandy soil, even by the sea, and its erect stems will slowly sucker to make a small thicket. The much-divided leaves turn to a bonfire of reds and orange in autumn, and leave behind reddish stems laden with scarlet hips. (See also page 25.)

The **rugosa roses,** too, grow and flower well on sandy soils and their large tomato-like hips last well into winter. ***Rosa rugosa* 'Alba'**♀ with glistening white simple flowers, which combine well with the showy hips, is a good selection.

***Mahonia aquifolium* 'Atropurpurea'** is a useful ground-covering shrub for poor soil and the foliage colours particularly well in winter, turning to rich shades of burgundy (see page 48).

Most **birches** relish a sandy soil and in winter their white bark is an illuminating sight rising up through a carpet of winter-flowering heathers or bronze sedges. ***Betula pendula***♀, the British native silver birch, has craggy silver-grey bark with dark crevices. It suits naturalistic plantings and looks good planted in groups. ***Betula pendula* 'Tristis'**♀ is perhaps the most elegant form, with a light airy habit and graceful weeping shoots. Although tall, it casts only light shade and is good in a confined space. (See also page 150.)

Betula pendula

Wet conditions

You may think it is hard to find winter-interest plants for wet and waterlogged soils, but look to the natural landscape. Willows will do especially well, and will offer rays of welcome colour with their bright young shoots. Some of the grasses and sedges can tolerate such conditions, provided they are not standing with their roots perpetually in water, and there is even a rush for the waterside.

If you have a good deal of space then consider the winter heliotrope, ***Petasites albus***. It needs to be planted with caution since it is a rapidly spreading colonizer, but it is an excellent ground-cover plant for a wild area in damp soil, even at the side of streams. Its winter feature is the scented white flower spikes emerging from the bare soil. These are followed by vast, jagged, kidney-shaped leaves, which are effective in seeing off brambles, nettles and any other competition. ***Petasites paradoxus*** is slightly less invasive, with scented flowers and white-backed leaves.

Few trees are more lovely in winter than ***Betula nigra***, the river birch, a fast-growing tree of light, airy form. Without its summer foliage its filigree winter silhouette is enhanced by peeling, shaggy pink-orange bark that catches the light and the breeze. ***Alnus incana* 'Aurea'** is a beautiful form of the grey alder with yellow foliage that falls away to reveal orange-yellow twigs, young bark and catkins, which glow in the winter sun.

Betula nigra

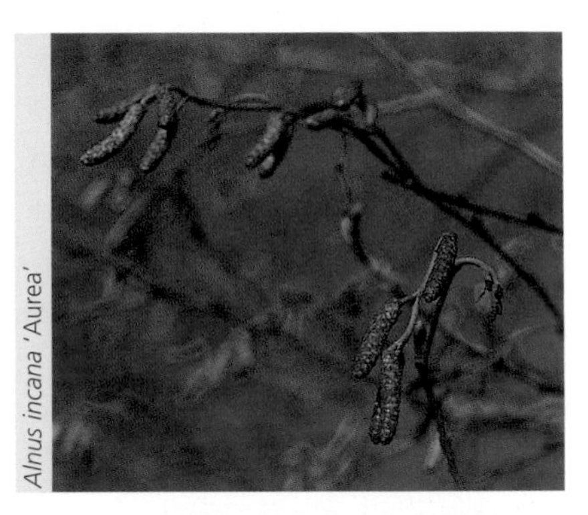
Alnus incana 'Aurea'

The willows (***Salix***) are deciduous shrubs and small trees that are very happy in wet soils (see also page 85). They often have strikingly coloured young stems, so are best planted in an open situation where these can be seen to greatest effect. ***Salix* 'Erythroflexuosa'** has colourful stems in burnished orange and brown and they are strangely twisted and contorted. It can grow into a vigorous tree if left unpruned and it makes a striking sight with its unusual branches; it can be cut to ground level in late spring each year to encourage new coloured stems. ***Salix babylonica* var. *pekinensis* 'Tortuosa'**🏆 also has stems that twist and turn, and these look especially dramatic when seen against a clear blue sky.

Salix 'Erythroflexuosa'

Petasites albus

MORE PLANTS FOR WET CONDITIONS *Alnus* × *spaethii* • *Andromeda polifolia* • *Chusquea culeou* •

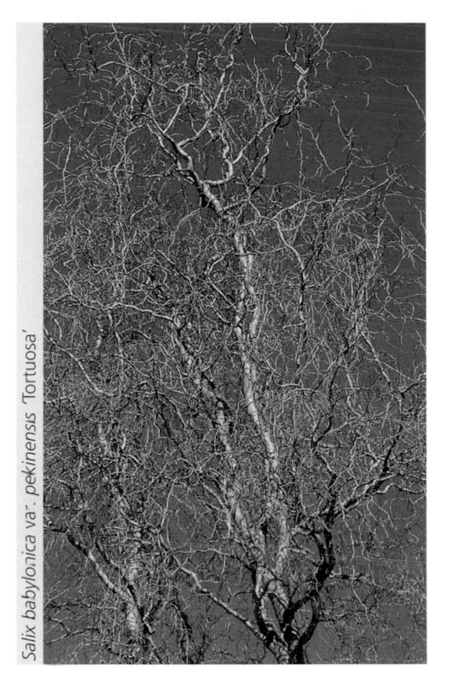
Salix babylonica var. *pekinensis* 'Tortuosa'

Salix alba var. *vitellina* 'Britzensis'

Some of the most colourful stems are on the *Salix alba* cultivars such as ***Salix alba* var. *vitellina***♀, with bright yellow new stems, and ***Salix alba* var. *vitellina* 'Britzensis'**♀, with orange scarlet stems. These willows grow best in deep, damp soil, and branches need to be cut back to the main trunk (pollarded) or stems cut right to the ground (coppiced) every other year to produce the most strongly coloured new growth.

The willows make excellent planting partners for the cultivars of ***Cornus alba*** and ***Cornus sericea*** that are also grown for their colourful winter stems and are equally happy in damp sites (see pages 147–48).

Many of the tufted hair grasses, the **deschampsias**, tolerate shade and poorly drained soil, growing into tussocks of dark green foliage 60cm (2ft) high, from which arise arching stems topped with airy spikelets of tiny flowers. ***Deschampsia cespitosa* 'Bronzeschleier'** (BRONZE VEIL) has bronze spikelets and ***Deschampsia cespitosa* 'Goldtau'** (GOLDEN DEW) has golden yellow flowers fading to buff. The leaves are almost evergreen and the flower stems remain through most of the winter. (See also page 31.)

The woodrushes (***Luzula***) too are happy in damp shady soil. The evergreen ***Luzula sylvatica*** makes good ground cover here, enlivened in the spring by its tassels of chestnut-brown flowers held well above the leaves. To bring winter colour into a shady spot plant small groups of ***Luzula sylvatica* 'Aurea'**, whose spring foliage is lime yellow turning yellowish green in summer and darkening to burnished gold in winter (see page 97). ***Luzula sylvatica* 'Taggart's Cream'** is a variegated form, with cream-striped new leaves. All grow to around 1m (3ft) when in flower.

The rush ***Juncus patens* 'Carman's Gray'** is excellent for wet conditions, doing well in very damp soil and even where its feet are in water. It forms an erect, steel-blue column of evergreen leaves up to 60cm (2ft) high, with cylindrical blue foliage and plenty of brown flowers.

CUT WILLOW STEMS

The bright stems of many willows look good in a winter arrangement for the house. They can also be mixed with other winter-foliage plants in containers outdoors, simply being pushed firmly into the compost; they may even produce new roots.

Deschampsia cespitosa 'Bronzeschleier' (BRONZE VEIL)

Cornus alba 'Sibirica' • *Cornus sericea* 'Budd's Yellow' • *Fraxinus excelsior* 'Jaspidea' • *Taxodium distichum* •

Woodland

Not all gardens have space for a wood but a similar environment can be created beneath just a few deciduous trees, a winding path of shredded bark providing access. In winter, light can penetrate the woodland floor, and there are lots of plants that take advantage of such conditions. If the lower branches of the trees are removed, to lift the canopy, small perennials will be able to grow right up to the trunks.

The evergreen rosettes of foxgloves, *Digitalis purpurea*, nestle among fallen oak leaves waiting for spring.

For evergreen structure and delicious perfume in a woodland setting, plant some of the ***Mahonia* × *media*** cultivars, with their stiff stems, long, spiny leaves and erect spikes of yellow flowers (see page 142). Introduce more scent with **sarcococcas** (Christmas box), which form compact bushes with neat foliage and tiny cream tassel flowers and will grow in quite deep shade (see page 143).

For rich winter foliage colour the **epimediums** are hard to beat. In the wild these low, ground-covering perennials grow in woodland and shady rocky places and, apart from the Japanese species, which require an acid soil, they thrive in most conditions. The species that hold their leaves until the spring, when the new foliage starts to appear, are the ones to grow, with the old leaves turning attractive shades of chestnut, mahogany and polished coppery red in the autumn. One of the toughest is ***Epimedium* × *rubrum***♕, with heart-shaped leaves that turn a rich copper in winter. ***Epimedium davidii*** comes in more unusual shades of ochre and brown, with more rounded, scalloped, prickle edges to the leaves. (See also page 55.) For extra colour plant a few Lenten roses, ***Helleborus* × *hybridus***, with flowers in white, cream, pale pink or wine red (see pages 72–74), and scatter around clumps of white snowdrops.

The spurge laurel, ***Daphne laureola***, is an uncomplaining, small bushy shrub that thrives in dry woodland conditions, even on chalk, with pleasant evergreen foliage

Mahonia × media 'Charity'

Sarcococca ruscifolia

Daphne laureola

OTHER GOOD PLANTS FOR WOODLAND *Galanthus nivalis* • *Iris foetidissima* • *Mahonia aquifolium* 'Apollo' •

and terminal clusters of little tubular yellow-green flowers peeping out from circlets of dark green leaves. Although not particularly showy, the flowers have a strong scent, most pronounced at dusk.

Plants with large striking leaves are needed in the woodland garden, since they reflect the dappled sunlight and show up among the leaf litter on the woodland floor. The wild ginger, ***Asarum europaeum***, is a good choice, with its glossy, rich green, kidney-shaped leaves making an evergreen carpet for spring-flowering phlox and tiarellas and autumn-flowering *Saxifraga fortunei*♕. The American species ***Asarum caudatum*** has larger, more rounded, matt green leaves and will even grow in coniferous woodland. Here too it is possible to establish colonies of the autumn-flowering ***Cyclamen hederifolium***♕, which likes being bone dry in summer; with a little moisture in winter, the pink or white flowers are followed by the marbled leaves (see page 65).

It is surprising that a green-flowered plant can stand out in the garden in midwinter, but this is the case with the stinking hellebore, ***Helleborus foetidus***♕, which flourishes in rough areas and in hedgerows and woodland in the wild. The very dark, holly-green, divided leaves grow in a fan shape from which arise stout stems, up to 60cm (2ft) high, topped by drooping clusters of bell-like, pale apple-green flowers, often rimmed with dark maroon. The flowers shine out from a dark corner and it is only at very close range that the unpleasant smell is noticeable. They are not long-lived plants but they produce plenty of seedlings to carry on the race. ***Helleborus foetidus* Wester Flisk Group** has striking claret-red stems and even more pronounced red edges to each little flower cup (see page 76).

A useful understorey carpeting plant is ***Pachysandra terminalis***, which produces whorls of evergreen leaves at the tips of short stems. Happy in all but limy soils, it can be used to cover the ground beneath rather gaunt shrubs such as *Mahonia* × *media* 'Charity'♕. There is a variegated form, ***Pachysandra terminalis* 'Variegata'**♕, with attractive rosettes of leaves outlined in cream.

The pretty blue-flowered periwinkles (***Vinca***) can be used in many situations, making good evergreen ground cover in shade; some cultivars have variegated leaves, and in some the flowers are reddish purple or white, as in ***Vinca minor* f. *alba***♕. They all need to be planted with caution otherwise they will soon envelop small perennials. One to cherish in sheltered woodland is ***Vinca difformis***♕, with little icy blue flowers appearing throughout the winter, tucked among the trails of fresh green leaves.

A really late-flowering woodland grass, which then lingers over winter, is ***Chasmanthium latifolium***. Making a clump 60cm (2ft) high with similar spread, it has broad green leaves and flattened oat-like flowerheads on drooping stems, taking on pinkish tones in cold weather.

Pachysandra terminalis

THE WOODLAND FLOOR

On the woodland floor there are a number of perennials that take advantage of the extra light and rainfall in winter when deciduous trees lose their leaves. In the garden these provide foliage interest and sometimes delicate flowers that thrive in winter shade.

Helleborus × *hybridus* 'Usha' (**1**), *Epimedium* × *rubrum*♕ (**2**), *Helleborus foetidus* Wester Flisk Group (**3**), *Asarum europaeum* (**4**), *Cyclamen hederifolium*♕ (**5**), and *Vinca minor* f. *alba*♕intermingling with *Galanthus nivalis*♕ (**6**).

Walls and fences

A warm sheltered wall is a cherished asset in any garden, particularly in winter, and all too easily filled to capacity, with plants growing at the base of the wall as well. For cold, shady walls and fences the choice is perhaps more limited, but there are still a good number of evergreen and deciduous shrubs and climbers that will thrive here, between them providing winter interest in the form of leaves or structure or really quite striking flowers.

Above: The cheery yellow flowers of *Jasminum nudiflorum* grace many a cottage wall.

Below: The leaves of *Cotoneaster horizontalis* colour red before they fall in autumn, leaving red berries shining on the branches.

COOL, SHADY SITES

Jasminum nudiflorum♀, winter jasmine, is perhaps the most popular winter-flowering plant for walls in sun or shade. Although naturally sprawling and untidy, it can be trimmed and trained. It can also be grown through other wall shrubs and climbers. (See page 82.)

A sunless wall beneath a window could be hard to fill but ***Cotoneaster horizontalis***♀ will thrive here, its low-growing herringbone branches spreading sideways to fill the allotted area. Its tiny dark green leaves are deciduous but colour well before they fall in autumn, to reveal the intriguing formation of the stems. In summer it bears small white flowers, followed by sealing-wax red berries, which persist well through the winter. This shrub needs no pruning: merely cut back any branches that overreach their space. The form ***Cotoneaster atropurpureus* 'Variegatus'**♀ is even better as the leaves have a neat cream outline, giving the whole plant a light silver effect in summer.

A patterned ivy (***Hedera***) is a natural choice for the challenging conditions of a cold, shady wall. There are many cultivars, some with large leaves and some with small, some vigorous growers and some more modest in their proportions. Leaf shapes and variegations are many, and by careful selection an ivy can be found to brighten and add interest to any dark corner or shady wall. (See pages 46–47, 93 and 98–99.)

Fatsia japonica♀, the Japanese aralia, is a useful shrub with big, glossy green, hand-shaped leaves carried on upright stems (see page 89). It is ideal against a shady wall and grows well in a pot at the base of a wall where there is no soil. Left unpruned it will reach 3m (10ft) or more. Old plants eventually become leggy with all the foliage at the top of the stems. These can be cut back hard in spring and vigorous new growth and healthy foliage will quickly follow. Fatsia has the bonus of large, white, ivy-like flowers in early winter, making a dramatic contrast against the dark green leaves.

The hybrid between ivy and fatsia, × ***Fatshedera lizei***♀, has large, ivy-shaped,

MORE EVERGREENS FOR SHADY WALLS *Azara microphylla* 'Variegata' • *Camellia japonica* • *Eriobotrya japonica* •

Garrya × *issaquahensis* 'Glasnevin Wine'

Pyracantha 'Orange Glow'

polished pale green leaves. In winter a tall snowdrop, such as *Galanthus* 'S. Arnott'🏆, will shine out planted at the base, with Solomon's seal (*Polygonatum*) to follow, intermingled with seedlings of Bowles' golden grass, *Milium effusum* 'Aureum'🏆.

A tall, upright shrub to grow against a shady wall is ***Garrya elliptica***, which was discovered by David Douglas in coastal forest in the north-west USA in 1828 and has proved a valuable winter-flowering plant ever since. It has large, leathery, crinkled leaves with a woolly undersurface, and in the depths of winter produces dangling clusters of palest creamy-green catkins. ***Garrya elliptica* 'James Roof'**🏆 has the largest catkins, up to 30cm (12in) long (see page 84). It is a vigorous grower but can be pruned after flowering to encourage new shoots from the base. The evergreen leaves can become rather dishevelled and unsightly in summer. There is a hybrid, ***Garrya* × *issaquahensis* 'Glasnevin Wine'**, with reddish-purple new growth. The accommodating *Geranium macrorrhizum* 'Album'🏆 could be grown at its feet to hide the base.

A shady house wall with no windows is ideal for the tall evergreen ***Rhamnus alaternus* 'Argenteovariegata'**🏆, with grey-green leaves neatly edged with cream. It has a narrow, upright habit of growth and works well with a contrasting plain green rounded shrub of ***Choisya ternata***🏆 in front, and a small bush of ***Daphne odora* 'Aureomarginata'** for scented flowers in winter (see pages 88 and 137).

The evergreen **pyracanthas** can be grown as free-standing shrubs and will reach 3m (10ft) or more, but their clusters of creamy-white flowers in spring and their red, orange or yellow berries are often better displayed when they are trained on a wall. They are fully hardy and will flower and fruit well even if grown against walls that get little sun. After flowering, the long new growths should be cut back to the developing fruit clusters. The birds seem to leave the yellow and orange berries until all other fruit has been taken, and a well-trained pyracantha covered with berries is a pleasing sight in winter. Recently varieties have been bred to resist the canker and fireblight that can disfigure many of the older cultivars. ***Pyracantha* 'Orange Glow'**🏆 is reliable, vigorous and very free-flowering, with masses of bright orange-red berries. Also recommended are ***Pyracantha*** DART'S RED **('Interrada')**🏆,

× *Fatshedera lizei* and *Hedera colchica* 'Sulphur Heart' clothe a shaded wall in evergreen foliage.

Euonymus fortunei 'Silver Queen' • *Hedera canariensis* 'Gloire de Marengo' • *Itea ilicifolia* • *Pileostegia viburnoides* •

Viburnum × burkwoodii 'Anne Russell'

Coronilla valentina subsp. *glauca* 'Citrina'

Chaenomeles × superba 'Pink Lady'

Drimys winteri

with large, matt red fruit, ***Pyracantha*** SAPHYR ORANGE (**'Cadange'**)♀, with deep orange berries, and ***Pyracantha rogersiana*** **'Flava'**♀, with bright yellow berries. ***Pyracantha*** **'Harlequin'** has very small, pale green, pink-flushed leaves splashed with cream at the edges; it is best grown as a foliage plant and is a good host for a clematis with light growth, such as *Clematis* ROSEMOOR ('Evipo002').

The hybrid ***Viburnum × burkwoodii*** **'Anne Russell'**♀ is an upright, evergreen shrub with pointed, oval leaves in dark green, with felted beige undersides. It can be grown against a shady wall, where the extremely fragrant clustered heads of pink-budded white flowers will show up well, opening from late winter onwards. The colourful evergreen **camellias** (see page 112) make lovely wall plants, but they need a site out of the morning sun, which would destroy any frosted blossom.

WARMER WALLS AND FENCES

The base of a dry, sunny wall or fence would be suitable for one of the **coronillas**, perhaps ***Coronilla valentina*** **subsp.** ***glauca*** **'Citrina'**♀, which grows to 1.5m (5ft), with pale lemon-yellow flowers, or the dwarf cultivar ***Coronilla valentina*** **subsp.** ***glauca*** **'Pygmaea'**, in bright yellow. These flaunt their scented pea-shaped flowers intermittently throughout winter, often continuing into summer, and the blue-green pinnate leaves are attractive in their own right. ***Coronilla valentina*** **subsp.** ***glauca*** **'Variegata'** has leaves edged in creamy white and looks delightful against a brick or stone wall. All need a bit of shelter but not rich feeding.

Although the ornamental quince (***Chaenomeles***) is tolerant of a shady position, if trained against a warm wall it will often flower in late winter. ***Chaenomeles × superba*** **'Pink Lady'**♀, with rose-pink flowers, darker in bud, is a reliable early-flowerer with plenty of blossom. Against a wall it will grow to 2m (6ft) with a similar spread.

Drimys winteri♀ (winter's bark) is an evergreen tree in its native Chile, but more often a large shrub in cultivation. It has large, shining, leathery leaves and bears clusters of fragrant ivory-white flowers on long stalks in early summer. The bark and twigs are aromatic. It prefers a moist climate and an acid soil. In colder areas it is best given wall protection, but when happy will prove a long-lived shrub.

ORNAMENTAL QUINCES

Chaenomeles are easy wall shrubs to grow, requiring only pruning and training after flowering in spring. Their glossy foliage and coppery new growth look good throughout spring and summer. Many produce aromatic, golden yellow, quince-like fruit in autumn, and their delicate blossom opens on bare stems from late winter onwards. When the buds have swollen on the branches, stems may be cut and brought into the house for early flowering in a vase.

Chaenomeles speciosa 'Nivalis' (above) Pure white flowers with golden stamens; lovely on a flint wall.

Chaenomeles speciosa 'Moerloosei'♀ Delicate pink and white flowers, resembling apple blossom.

Chaenomeles speciosa 'Geisha Girl'♀ Double, deep apricot-pink flowers.

Chaenomeles × *superba* 'Crimson and Gold'♀ Deep crimson petals, golden anthers; very showy.

MORE EVERGREENS FOR SUNNY WALLS *Ceanothus arboreus* 'Trewithen Blue' • *Daphne bholua* 'Jacqueline Postill' •

CLIMBERS FOR WARM WALLS

Winter is not the season for flowering climbers, although there are some clematis that excel at this time of year: their delicate blooms are cherished by gardeners and they have become among the most popular of garden plants. Although evergreen climbers are normally associated with shady walls, the domain of the ivies, there are those that prefer a sunnier position.

The finest winter-flowering clematis is *Clematis cirrhosa* var. *balearica* (1), a Mediterranean evergreen, growing up to 4m (13ft). It is known as the fern-leaved clematis for the pretty, finely cut, glossy leaves, which are divided into leaflets and take on bronzy tints in winter. The small bell-shaped flowers flare out at the tips and are pale cream with brownish-red speckles all over the inner surface. They are pleasantly lemony scented and appear in clusters throughout the winter. The flowers can turn pink as they age and they dry to form silky seedheads. The scent is most pronounced if a few sprigs are cut and brought indoors.

This clematis can be grown into a tree, where the hanging creamy bells look delightful, but it is more usually trained against a wall, preferably with a light background to provide a contrast for the dark green, ferny leaves. It is reasonably hardy but benefits from the protection of a warm wall and occasionally needs to be cut back hard after flowering to keep it in shape. In colder areas the old leaves may die before the new foliage appears in spring and this is the time to do any pruning.

The cultivar aptly named *Clematis cirrhosa* var. *purpurascens* 'Freckles'🏆 (2) is autumn-flowering, with the insides of the petals heavily marked with maroon speckles, while *Clematis cirrhosa* 'Wisley Cream'🏆 (3) has larger, unspotted cream flowers and less divided leaves.

The lovely winter-flowering *Clematis napaulensis* bears unusual creamy-yellow, nodding, bell-shaped flowers, with the petals recurving at the tips and prominent stamens with purple anthers protruding from each flower. Growing to 4m (13ft), this is quite a vigorous climber, with bright green leaves, becoming dormant in summer before leafing again in the autumn. It must have the shelter of a warm, sunny wall or a conservatory.

It is difficult to recommend the other evergreen clematis, *Clematis armandii*, since the large leathery leaves become very drab and tatty by the end of winter and remain so until they are shed in summer. The creamy flowers appear in early spring.

Trachelospermum jasminoides🏆 (4) is a twining climber, needing wires or trellis to haul itself up, but it will then smother a wall with narrow, pointed, dark evergreen leaves, which catch the sun in winter. In summer it has scented ivory-white jasmine-like flowers. It is considered slightly tender, but seems to succeed on a warm sunny wall in all but the coldest of years. The variegated form *Trachelospermum jasminoides* 'Variegatum'🏆 is proving more robust and is a lovely climber, even if rather slow-growing. In colder areas the foliage of all trachelospermums flushes crimson-purple in winter, the variegated form taking on lovely pinkish hues.

Hardier than *Trachelospermum jasminoides, Trachelospermum asiaticum*🏆 (5) is also an evergreen twiner, with glossy green foliage and beautifully scented, ivory to creamy-yellow flowers in summer. It is a suitable host for a viticella clematis, which will give later flowers.

Euonymus fortunei 'Emerald 'n' Gold' • *Hydrangea serratifolia* • *Magnolia grandiflora* • *Sophora* SUN KING ('Hilsop') •

Pots and containers

A pleasing winter scene can be created using plants in pots and containers so that even gardeners with limited space can enjoy winter flowers and foliage – and the beauty of containers is that the picture can easily be changed as the season progresses. Place pots of winter plants to bring new life to fading borders, or group them near the house so that they can be viewed from the windows.

A large clay pot, wooden half-barrel, stone urn or metal container can be planted up with several winter-interest plants. A formal effect can be achieved with evergreen foliage such as box trained in topiary shapes, or a really large terracotta jar or glazed pot can be left empty and placed in a strategic spot. Although single pots containing several plants can be very attractive, an even better effect can be achieved with groups of containers of varying sizes; this is especially so if the pots used are all of a kind, such as terracotta

Above: A varied and attractive planting picture can be created by grouping pots and containers together. Small plants are not lost in isolation; pots can be added or taken away as the season advances.

Right: A terracotta pot of thyme looks good throughout winter; the leaves can be picked for cooking.

or contemporary galvanized metal, or are glazed in similar colours, then the plants – not the pots – will take centre stage. Use a mixture of foliage plants in different colours, adding pots of flowering plants as they become available, and you will have a flourishing small garden all through the

MORE FLOWERS FOR WINTER POTS *Chionodoxa luciliae* • *Crocus chrysanthus* 'Ladykiller' • *Cyclamen* Miracle series •

CARING FOR PLANTS IN POTS

A wide range of plants are amenable to growing all year in containers, but they do need a certain amount of care, with particular attention paid to watering. It is important that they receive enough water in summer since the compost will dry out extremely quickly in hot weather, especially in small pots. Conversely, in winter, if overwatered, some plants can rot in compost that is too wet. Stand the pots on a porous surface such as gravel and make sure that water can drain away from the base – large pots will need to be raised from the ground with bricks or special pot feet. Some evergreens, if kept in too small a container, can die of desiccation in winter, because the roots can freeze and prevent the plant taking up water. It is a good idea to group pots together so that they may provide mutual protection.

Use a compost that is suitable for containers, preferably a loam-based compost with some form of slow-release fertilizer added if the plants are to remain in the pots for a long time. It is not necessary to place broken crocks in the bottom of the pots, but it would be wise to mix grit in with a very light compost, particularly the peat-free kind, to add weight and so help prevent the pots falling over.

In time the surface of the compost may become covered with algae or moss: a top layer of gravel or crushed bark will help to prevent this and will look attractive. Plants in small pots can be repotted in the spring, using fresh compost; those in larger containers need to have the top layer of compost scraped off and replaced.

winter. A few trailing plants such as ivy can help to soften the edge of a container.

A pleasing group for a sunny spot could comprise a large pot of the shrubby ***Skimmia japonica*** **'Rubella'**🏆, with its deep green leaves and glowing ruby buds; the grassy leaves of ***Anemanthele lessoniana*** (formerly *Stipa arundinacea*), forming a waterfall of orange and red in winter; pink stems and silver-striped leaves on ***Euphorbia characias*** SILVER SWAN (**'Wilcott'**); and the crinkled maroon leaves of ***Heuchera*** **'Plum Pudding'**. All these could be set off by silver rosettes of ***Lychnis coronaria***🏆 and the plain green, locket-shaped leaves of an **epimedium**. Add a large pot of the evergreen ***Liriope muscari*** **'Variegata'**, with broad, grassy leaves edged with pale gold, to complete the picture (see left above). (See also pages 30, 57 and 59.) The skimmia will need to be moved out of the sun in summer.

In early winter, pots of temporary colour can be added using dwarf half-hardy cyclamen, early hybrid primroses and heathers (see page 83). Winter pansies come in a wide range of bright colours and will flower spasmodically all season, as long as they are regularly deadheaded, although they are at their best in late winter and early spring. Garden centres also offer a wide range of dwarf evergreens grown for winter bedding and these are useful fillers. From midwinter onwards pots of early bulbs such as dwarf iris, snowdrops, crocuses and *Cyclamen coum*🏆 continue the colour into spring.

For a grouping that will do well in shade, consider ***Skimmia* × *confusa*** **'Kew Green'**🏆, whose apple-green foliage and pale cream flower buds (see page 49) would stand out from the

Right (from top to bottom): Snowdrops and aconites with *Ajuga reptans* 'Atropurpurea'; a clump of snowdrops, *Galanthus nivalis*, lifted from the garden and potted for the terrace; and *Helleborus niger*, the Christmas rose.

Erica carnea 'Springwood White' • *Iris* 'George' • *Iris danfordiae* • *Primula* Cowichan series • *Scilla siberica* •

Evergreen perennials combine here with seasonal colour provided by cyclamen and violas. The dark foliage of *Ophiopogon planiscapus* 'Nigrescens' (top left) and *Euphorbia* BLACKBIRD (bottom right) is a dramatic contrast to lighter foliage, feathery grasses and delicate flowers.

shadows, accompanied by ferns such as the evergreen ***Asplenium scolopendrium* 'Kaye's Lacerated'**♀, with broad, bright green fronds, deeply cut and waved at the edges, and the feathery leaves of ***Dryopteris erythrosora***♀ (see page 91). Although the bright lime-green foliage of ***Heuchera*** KEY LIME PIE (**'Tnheu042'**) fades in winter, this paler colour shines out in shade, and the softly ruffled leaves add texture (see page 59). To complete this grouping drop in pots of snowdrops when they start to flower and containers of the little golden feverfew, ***Tanacetum parthenium* 'Aureum'**, which always provides a few seedlings for this purpose. Variegated ivies could be allowed to weave through the other plants. Many combinations of these plants could be placed in a single large container if there is not enough space for a group.

For rich foliage colour there is a wealth of material to choose from. In recent years more purple-leaved plants have appeared and there are countless **heucheras** in tones of purple, metallic silver and almost black. ***Heuchera*** LICORICE (**'Tnheu044'**), **'Can-can'**♀, **'Cascade Dawn'** and **'Prince'** are all good in winter. The small purple **phormiums** offer a contrasting shape – purple in ***Phormium* 'Dark Delight'**, bronzed purple in **'Surfer Bronze'**, crimson and bronze in ***Phormium* 'Maori Chief'** and pink in ***Phormium* 'Maori Sunrise'**. The newer ***Euphorbia*** BLACKBIRD (**'Nothowlee'**) has a bushy habit and narrow leaves in shades of grape purple and maroon, and the mahogany-leaved sedge ***Uncinia rubra*** has stiff

MORE EVERGREENS FOR WINTER POTS *Gaultheria procumbens* • *Hebe* 'Emerald Gem' • *Leucothoe* SCARLETTA •

The spiky form of *Chamaerops humilis* var. *argentea* contrasts with the soft mounds of heucheras and carex (top). To add interest to a group of pots use plants with a variety of foliage shapes and textures, such as *Ilex cornuta* (left), *Astelia chathamica* (above left) and *Buxus sempervirens* 'Elegantissima' (above right).

grassy foliage. Further leaf colours come in shades of tawny gold and apricot in ***Heuchera*** CRÈME BRÛLÉ (**'Tnheu041'**). (See also pages 31, 52–53 and 58–59.)

Some short-lived colour combinations can be achieved using the cut stems of dogwoods and willows pushed into pots of winter plants. ***Cornus alba*** **'Sibirica'**♕, with its ruby-red stems (see page 111), combines well with the bronze threads of ***Carex buchananii***♕ (see page 101). This is a good choice for a pot because it has an upright habit rather than the more sprawling form of the other brown New Zealand sedges. Alternatively, place the cornus stems with ***Helleborus foetidus***♕ to echo the red rims to the clusters of bell-like pale green flowers (see page 123).

The large Lenten roses, ***Helleborus* × *hybridus***, are not particularly suited to containers as they have an extensive root system that is not easy to confine, but some ***Helleborus* × *sternii*** selections are much smaller plants (see page 76). Their attractively patterned leaves are swirled in silver, cream and green, the leaf stalks are red and the flowers are a pale creamy pink. If planted in deep 'long tom' pots and fed in the spring, they will be happy for several years.

THE SENSORY GARDEN

Beautiful flowers and colourful foliage are not the only aspects to consider when choosing plants for the garden in winter. Scent makes a big impact at this time of year, and texture can be felt in the bark of trees and shrubs. Vegetables and herbs can offer fresh taste, and movement is appreciated in the form of trembling foliage and swaying grasses and bamboos. Sunlight in winter has a particular quality, often illuminating shapes that are not apparent at other seasons.

RIGHT: Grasses and sedum heads catch the winter light.

Light

The changing light is one of the most wonderful aspects of winter. Bright and glittering when skies are clear, it can soon become dark and brooding with the threat of rain or snow. In the early morning mist, shapes and structures tend to merge, defined only by glistening cobwebs strung from every stem and branch. There are flashes of jewel-like colour from frosted flowers and berries, and shafts of glowing pastel as soft sunlight catches the coloured shoots of trees and shrubs.

Helleborus orientalis

Late afternoon sun illuminates the dried plumes of *Miscanthus sinensis*.

The light in the garden in winter is quite different from the sunny brightness of summer. In temperate latitudes the angle of the sun in winter is low and it picks up and reflects light on glossy leaves and shining bark. In open sunny aspects light will bounce off large-leaved evergreen shrubs: laurels and hollies are especially effective (see page 88), and the small undulating leaves of green pittosporums also offer a shiny, reflective surface (see page 92).

In contrast, some parts of the garden will be in deeper shade in the winter, as the sun fails to get over a wall or fence, or reach into a corner. Many plants can cope with just such a situation, provided the area is not overhung by conifers or evergreen branches. Flowers with light or bright petals are a good choice here: snowdrops, aconites and winter cyclamen all flower and finish their growth cycle very early in the year, before summer growers cast them further into shade (see pages

PLANNING FOR WINTER

When designing a garden with winter in mind, take careful consideration of the views from the house windows (see pages 12–13). Note where the sun strikes and where the strongest shadows form, and remember that new aspects will be opened up when the leaves fall in the autumn.

64–65 and pages 70–71). The winter-flowering honeysuckles, daphnes and viburnums fill the air with rich fragrance and their delicate flowers show up well against a dark background (see pages 137 and 142–43).

The low light enhances the colour in dried grass foliage, highlighting the fawn, biscuit and parchment leaves. The dried flowerheads of miscanthus ignite when backlit by winter sun, and the fine curling leaves of carex move and flicker like wispy flames rising from the soil. The sun also picks up the rusty tones of many seedheads. Brown sedum flowerheads on bleached stems turn chestnut in the morning or evening sun. (See pages 29–31.)

Deciduous trees take on a peculiar majesty against the winter sky, particularly on clear evenings in the glow of the setting sun. The dark skeletons of oak and ash seem to grow in stature before nightfall, and cedars become dense inky silhouettes. In the garden the definite shapes of dark structure plants like yew and holly contrast with the light, airy tracery of birch branches rising high above them.

Late afternoon on sunny days is the time to warm the soul admiring the stems of dogwood, willow and bamboo. Plant these where their rich hues will catch the yellow light of approaching evening. This is also the time to enjoy trees with polished brown bark such as the Tibetan cherry, *Prunus serrula*; white-barked birches become more apparent as the light fades. (See pages 144–46.)

There is no denying that some winter days are unrelentingly grey and, for these, there must be a powerful incentive to take a walk in the garden: small splashes of brilliant colour are what is needed. Winter-flowering heathers, *Cyclamen coum*, *Daphne mezereum*, early rhododendrons, camellias and hellebores will give this sparkle to the winter garden. (See pages 65, 72–77, 83 and 112–14.)

If rain has set in, then the garden will be looked at from within the house so eye-catching plants need to be visible from every window. In early winter there is great pleasure in watching a patch of bergenias slowly changing in colour from bright moss green to burnished bronze, copper and finally rhubarb red (see pages 54–55). The first welcome flowers on the winter jasmine, *Jasminum nudiflorum*, offer rays of sunshine on a grey day (see page 82).

The garden is transformed after an overnight frost, the leaves and stems outlined in silver and seeming brittle in the still air. Frosty days may be clear and blue or hazy and magical. If snow does fall, then the light changes again, usually leaden and grey before the snow comes and then, when it has settled and the sun comes out, all is brightness and sparkle set against a blue sky.

Above: The brilliance of the winter morning sun is accentuated by a light fall of snow.

Below: The afternoon sun shines mellow and warm through the copper leaves of a beech hedge.

Scent

Shrubs with fragrant flowers have always been valued in our gardens, particularly in winter. In this season there are fewer insects and other pollinators about so certain plants send out a strong perfume that carries far and wide; they also tend to continue blooming far longer than their summer cousins. Foliage is often scented, too, especially that of conifers like junipers and thujas and some herbs. Some of the winter-flowering bulbs are fragrant, although their perfumes are more elusive, but the few perennials that are in bloom now seem to rely on the visual attraction of their flowers alone (see pages 72–79).

The fragile flowers of witch hazel (*Hamamelis*) catch the light on a cold morning; their spicy fragrance hangs on the cold air.

Commonly known as the white forsythia, ***Abeliophyllum distichum*** is a light airy shrub that will grow to 2m (6ft) in height but is usually seen as a much smaller plant. Palest lilac buds open white all along the bare purplish stems in late winter. Although hardy it needs a sunny site in order to flower well and to bring out the almond perfume. It looks enchanting with a pink winter-flowering heather such as *Erica carnea* 'Winter Beauty' in a carpet at its feet.

The cascading yellow flowers of ***Azara microphylla***♡ are not particularly large, but their delicious vanilla fragrance is surprisingly strong, especially on a mild day in late winter. This small-leaved evergreen from South America makes a small tree but it is quite narrow-growing, even though its branches arch, so it can be accommodated in a bed at the base of a warm wall.

The wintersweet, ***Chimonanthus praecox***, has been grown in gardens since the 18th century, when it was introduced from China, and it is treasured for its sweetly scented flowers on leafless branches in winter. It can make a large shrub but it is usually trained against a warm wall to encourage it to develop good buds for the next season's flowers. These starry flowers have irregular, narrow petals, which vary in colour from translucent ivory and cream to deep yellow, often with the shorter inner petals marked with maroon at the base. *Chimonanthus praecox* 'Grandiflorus'♡ is a better form, with larger, yellow flowers and a conspicuous purple stain

SITING SCENTED SHRUBS

Many of these shrubs are best planted at the back of borders where their flowers and fragrance can be appreciated in winter and then they can slip unobtrusively into the background in summer. You may also want to place one or two scented plants with more attractive foliage near a gateway or beside a frequented path so that you can appreciate their perfume as you pass by.

Abeliophyllum distichum

Chimonanthus praecox 'Luteus'

Daphne bholua

at the base of the petals. ***Chimonanthus praecox* 'Luteus'**♀ has large, more regular, unmarked flowers of golden yellow, strongly scented of exotic spices. Unfortunately young plants take a while to settle in and start producing flowers, so patience is needed. Wintersweet has undistinguished leaves in the summer, but can act as host for a clematis.

A **daphne** bush in full scented flower is one of the wonders of winter, stopping people in their tracks. Some winter-fragrant plants have inconspicuous flowers and you wonder where the scent is coming from, but a bush of any of the ***Daphne bholua*** forms smothered in tight clusters of pink or purple-backed white flowers cannot be missed. *Daphne bholua* (*bholua* being the native Nepalese name for the plant) is a tall, upright shrub and not as hardy as some of the named cultivars, which make better garden plants. These usually flower from midwinter onwards, before the leaves appear, although some can be evergreen. ***Daphne bholua* var. *glacialis* 'Gurkha'** is particularly hardy and well scented. It was originally collected at high altitude in eastern Nepal and is a deciduous form, producing its purple-pink and white flowers on the bare wood. ***Daphne***

Daphne bholua 'Jacqueline Postill'

***bholua* 'Jacqueline Postill'**♀ has large, showy, strongly scented flowers, which are purplish pink in bud opening creamy pink. It was raised by Hillier propagator Alan Postill in 1982 and has proved to be a very hardy and excellent garden plant. It is normally evergreen but will shed some of its leaves in a cold exposed position. Others worth considering are ***Daphne bholua* 'Peter Smithers'**, smelling of lemon and jasmine, and the early-flowering ***Daphne bholua* 'Darjeeling'**. Both are evergreen or semi-evergreen.

The evergreen ***Daphne odora*** has the richest scent of all the winter-flowering daphnes. Although the species is considered tender, the forms with gold rims to the edges of the dark green leaves have proved hardier. ***Daphne odora* 'Aureomarginata'**♀ is most commonly grown. Older plants often become infected by a virus, which causes the foliage to become mottled. Eventually the plants dwindle and die, and the only solution is to replace them. The average life expectancy of *Daphne odora* 'Aureomarginata' is ten years or so, and it is advisable to have a replacement plant ready to take over from an elderly specimen; happily, it is relatively

Daphne odora 'Aureomarginata'

easy to propagate from semi-hardwood cuttings taken in summer.

There is an unusual deciduous daphne, ***Daphne jezoensis***, with scented yellow flowers in winter, borne on a small bush that produces its new leaves in autumn and sheds them in summer. For descriptions of ***Daphne mezereum*** and ***Daphne laureola*** see pages 117 and 122.

Daphne jezoensis

The coppery orange flowers of *Hamamelis* × *intermedia* 'Aphrodite' (see page 141) glow in the afternoon sun; in the background, the parchment plumes of miscanthus rise above the planting.

All daphnes thrive in either sun or partial shade and will grow on acid or alkaline soil as long as it is moist but well-drained: they need constant moisture throughout the year, disliking it either too wet or too dry. They generally do not grow well in pots, faring better in open ground. They are ideally placed close to the house, where the fragrance can be appreciated as you enter or leave.

Edgeworthias are closely related to daphnes and, like them, bear richly scented flowers in winter. They were named for Michael Edgeworth (1812–1881), an amateur botanist who collected many new plants while working for the civil service in India. ***Edgeworthia chrysantha*** grows to 1.2m (4ft) high and forms a regularly branched bush with upright shoots. Silky buds develop in clusters at the tips of the branches in early winter, opening into sweetly perfumed, rich yellow flowers, covered with shining white hairs. In the variety ***Edgeworthia chrysantha* 'Red Dragon'** they are a startling bright orange. Edgeworthias are best grown in light dappled shade in a sheltered position. They are usually deciduous, producing new leaves after flowering. They have strangely pliable wood that in Japan is used in the manufacture of paper for bank notes.

Edgeworthia chrysantha

Edgeworthia chrysantha 'Red Dragon'

The evergreen ***Elaeagnus* × *ebbingei*** and ***Elaeagnus pungens*** come into flower in late autumn and early winter, with such tiny cream flowers tucked beneath the leaves that it is often almost impossible to identify the source of the perfume. These are large backbone shrubs for a border, able to survive dry conditions and amenable to being pruned; they are also suitable as a hedge. The silvery underside to the leaves adds a lightening effect, and there are many selections with variegated foliage (see page 99).

(continued on page 142)

SCENTED FOLIAGE

Many plants have foliage that is aromatic when crushed or brushed against, and although the scent is not particularly noticeable in winter, it is nonetheless pleasant to rub the leaves of these plants when passing. Santolinas (**1**), the cotton lavenders, have a pleasingly astringent scent when crushed, and the smell of true lavender (**2**) is very distinctive, while the thymes (**3**) are variously scented of lemon, caraway and eucalyptus, as well as traditional thyme.

The alpine mint bush, *Prostanthera cuneata*♀ (**4**), forms a small evergreen shrub, with bright emerald-green leaves arranged in tight little rosettes all along the stems. They have a fresh aromatic smell reminiscent of eucalyptus when crushed, and in summer pale lilac flowers perch on the end of the stems like small butterflies. This is the hardiest form of prostanthera, growing high up in the Snowy Mountains in eastern Australia, and survives -10°C (14°F) if planted against a warm wall in full sun.

The upright, bushy evergreen *Myrtus communis*♀, the common myrtle, has small, glossy, dark green leaves smelling of juniper, which set off its pretty white summer flowers (**5**). Widely planted in the Mediterranean, myrtle was introduced into Britain in the 16th century and was considered to bring good luck to the household if planted by a doorway; small sprays of myrtle were also traditionally placed in bridal bouquets. The more compact form *Myrtus communis* subsp. *tarentina*♀ is a better choice for a small garden and has pink-tinged flowers in the autumn. There are forms with cream margins to the leaves, but these generally seem to be less hardy, although they are good seaside plants.

Rhododendron 'Praecox'♀ (see page 85) has aromatic leaves when crushed, as does the Mexican orange blossom, *Choisya ternata*♀ (**6**) (see page 88). *Skimmia anquetilia* and *Skimmia* × *confusa* (**7**) are fine winter-berrying evergreen shrubs that also have aromatic foliage when crushed, although the smell is rather pungent and not altogether pleasant (see page 49). Many of the conifers have spicy scents, with some thujas (**8**) smelling like the shavings from sharpened pencils (see page 51).

Some scented foliage is definitely unpleasant. Some people dislike the smell of box plants, and the evergreen *Iris foetidissima*♀, commonly known as stinking iris (see page 61), truly lives up to its name, having leaves that smell of roast beef (or rotting meat), presumably to attract pollinating flies.

WITCH HAZELS

The witch hazels (*Hamamelis*) have unusually shaped flowers, curling like little ribbons on the bare branches, defying winter weather and lasting for weeks. The narrow, spider-like petals in gold, ruby, amber or flame are held in maroon calyces and they emit a powerful spicy scent that can fill the garden on a still winter's day. Witch hazels form large shrubs or small trees with a spreading branch pattern and should not be pruned as this will spoil the shape. They are best planted in isolation against an evergreen background where the scent can waft freely, and they are happiest and flower most profusely on a neutral or acid soil.

Hamamelis mollis♀ (**1**), the Chinese witch hazel, has rich yellow flowers and is said to have the strongest scent of the witch hazels; the leaves turn butter yellow in the autumn. It grows into a large open shrub and is a good choice for a naturalistic setting.

Hamamelis 'Brevipetala' (**2**) is an upright form with thick clusters of short-petalled, deep yellow flowers appearing orange when seen from a distance. They crowd the branches in midwinter, giving the impression that the whole branch is glowing with tiny flames. The scent is heavy and sweet.

The hybrids grouped under *Hamamelis × intermedia* come in a variety of colours, from palest primrose-yellow in 'Pallida'♀ (**3** and see Good Companions, right), to rich brick red in 'Diane'♀ (**4**). 'Jelena'♀ (**5** and see Good Companions, right) is a superb vigorous variety with yellow flowers suffused with coppery red; the leaves turn scarlet before falling. 'Primavera' is more upright and has yellow petals tinged with purple. 'Harry' (**6**) has large orange flowers, darker red at the base of the petals. A single flower cluster resembles the centre of a succulent peach with the stone ripped from it. 'Arnold Promise'♀ is a vigorous variety of wide-spreading habit with attractively pleated leaves. The bright yellow flowers open in late winter. 'Aphrodite' (**7**) is another spreading shrub with large deep orange flowers and good yellow autumn foliage colour. Both copper-red 'Ruby Glow' and light orange 'Orange Peel' (**8**) are more upright in habit with rich orange-red autumn colour.

The forms with the palest flowers show up best against a grey sky. The red- and orange-flowered cultivars are exquisite when backlit by the morning or late afternoon sun.

Hamamelis vernalis (**9**), the Ozark witch hazel, an American native, is worthy of mention. A tall, upright shrub, it has smaller flowers than other witch hazels but they are produced in great profusion. Usually pale orange to copper in colour, they have a heavy pungent scent, quite unlike that of the Chinese witch hazels. *Hamamelis vernalis* 'Sandra'♀, with strong yellow flowers, is the finest cultivar, raised at Hillier Nurseries. The foliage is most attractive: young leaves are flushed purple, and the autumn colour is a spectacular cocktail of orange, flame and red.

BUYING WITCH HAZELS

The Sir Harold Hillier Gardens in Hampshire hold the National Collection of the genus *Hamamelis*. This is a wonderful place to see young and mature witch hazels and to choose favourites. There are many varieties and they are certainly best selected when in bloom so that you can see at first hand the flower colour and shape and make sure that it is scented. Witch hazels are expensive to buy because they are grafted and are tricky to propagate. Occasionally seed-raised witch hazels are offered, but these can be disappointing in flower and lack perfume.

WHY WITCH HAZEL?

The common name witch hazel comes from the North American species *Hamamelis virginiana*. Early settlers cut branches of the shrub and used it for water divining in the same way as they used branches of hazel in England. The leaves and wood were similar, and the branches had the same magical properties in locating water, hence it earned its name. *Hamamelis virginiana* is also the commercial source of the witch hazel astringent.

GOOD COMPANIONS

The yellow-flowered witch hazel *Hamamelis* × *intermedia* 'Pallida'♀ is set off well by the leaves of the evergreen pampas grass *Cortaderia selloana* 'Aureolineata'♀.

Hamamelis × *intermedia* 'Jelena'♀ (1) looks stunning rising above the white stems of *Rubus cockburnianus* 'Goldenvale'♀ and the black grassy leaves of *Ophiopogon planiscapus* 'Nigrescens'♀ (2).

Mahonia japonica

Mahonia × media 'Winter Sun'

One of the earliest of the Asian **mahonias** to be introduced, ***Mahonia japonica***🏆 is a large, upright evergreen shrub with softly spined, pinnate sea-green leaves and arching sprays of pale primrose-yellow flowers, smelling strongly of lily-of-the-valley. This is still one of the finest species to grow, being happy in shade and almost any type of soil and flowering profusely. The first flowers appear in early winter and a few are still being produced as the season draws to a close. If grown in full sun and poor soil some of the leaves will turn attractive shades of orange and scarlet in winter.

Mahonia japonica has been crossed with the spectacularly flowered, but less hardy, *Mahonia lomariifolia*🏆 to produce ***Mahonia* × *media***, and there are now a number of popular cultivars. These architectural shrubs make substantial bushes up to 2.5m (8ft) high, with long leathery, evergreen pinnate leaves, arranged in whorls round a stiff stem, and bearing upright clusters of soft yellow flowers. ***Mahonia* × *media* 'Lionel Fortescue'**🏆 is the first to bloom in late autumn, followed by ***Mahonia* × *media* 'Winter Sun'**🏆, with brilliant yellow flowers, and ***Mahonia* × *media* 'Underway'**🏆, a compact form with long bright yellow racemes, then the popular **'Charity'**, with more drooping flowers, not quite so strongly scented. Strings of greenish-plum fruit follow the flowers.

These mahonias are useful plants for light woodland, where they provide pale evergreen bulk among deciduous shrubs; plant them near a path so that the scent can be appreciated – but not too near because of the prickly leaves. In time the plants can become large and ungainly and they benefit from having individual stems cut to the ground in spring, when they will rapidly produce new shoots.

There are several winter-flowering shrubby **honeysuckles**, all with strongly scented flowers. The bushes are unremarkable in summer, so they are ideal at the back of the border, against a fence or shed or even as part of an informal hedge or screen. A hybrid between two fragrant winter-flowering species with double the scent is ***Lonicera* × *purpusii* 'Winter Beauty'**🏆, bearing little crumpled flowers in pale buttermilk yellow on long bare branches intermittently from midwinter to early spring. It forms a rather sprawling bush, which can be pruned after flowering to keep it within bounds. In mild winters this honeysuckle can retain its leaves, which is a disadvantage as they tend to obscure the flowers.

Lonicera × purpusii 'Winter Beauty'

MORE PLANTS WITH FRAGRANT WINTER FLOWERS *Acacia dealbata* • *Camellia sasanqua* • *Erica × darleyensis* •

Sarcococca hookeriana var. *digyna*

Viburnum × *bodnantense* 'Dawn'

One of the strongest scents in midwinter comes from **sarcococca** (Christmas box). In fact, planted en masse it can be overpowering. The first to flower is ***Sarcococca confusa***♡, with neat, pointed, evergreen leaves on a compact bush up to 1.2m (4ft) high and as much across. The cream, tassel-like flowers are clasped along the stems and they are followed by black berries. ***Sarcococca ruscifolia*** (see page 122) is similar but with thicker dark green leaves and red berries. The most attractive species is ***Sarcococca hookeriana* var. *digyna***♡, which has a suckering habit and narrow mid-green leaves with reddish stems; its flowers are larger than the others, with pink on the backs of the petals, and are equally well scented. The cultivar **'Purple Stem'** has particularly fine purple stems and leaf-stalks, and even the leaf midribs are flushed with purple. These tough hardy plants come from Asia and are happy in shade, even dry shade, and grow on any soil. They need an unobtrusive spot near the house or a pathway for the scent to be appreciated and are a good choice for a shady corner where nothing else will grow, even managing under conifers.

When choosing a deciduous **viburnum**, one of the best for frost-resistant winter flowers is ***Viburnum* × *bodnantense***. The most popular variety is ***Viburnum* × *bodnantense* 'Dawn'**♡, which was bred in the 1930s. It grows into a stiff upright shrub to about 3m (10ft) tall, and bears strongly scented, dark pink flowers in large clusters on the bare stems. ***Viburnum* × *bodnantense* 'Charles Lamont'**♡, with flowers in paler pink, and the almost white ***Viburnum* × *bodnantense* 'Deben'**♡ are also well worth considering.

One of the parents of *Viburnum* × *bodnantense* 'Dawn' is ***Viburnum farreri***♡, an old favourite with a strong perfume, reminiscent of almond paste, and this scent travels well. It used to be known as *Viburnum fragrans* but was renamed to commemorate the great plant hunter Reginald Farrer, who introduced the species from China (although it had been discovered by William Purdom). The small clusters of little tubular flowers are pink in bud opening to almost pure white and they appear intermittently throughout the winter. The new leaves emerge bronze-tipped and they fade in the autumn in a flurry of purple.

Viburnum × *bodnantense* 'Charles Lamont'

PRUNING VIBURNUMS

These deciduous viburnums are tolerant of most soils except very wet conditions. They are vigorous shrubs and often get rather large for their location. They should be firmly pruned in the spring to remove old growth and promote good flowering. Cutting out a few branches right to the base of the plant is a better approach than pruning to half their length, which results in multiple straight shoots that detract from the appearance of the plant.

Lonicera fragrantissima • *Prunus mume* 'Beni-chidori' • *Skimmia japonica* 'Fragrans' • *Viburnum farreri* 'Candidissimum' •

Texture

Patterned bark and gleaming stems bring a variety of textures to the garden at any time of year but their beauty is more apparent in winter as leaves are shed and herbaceous plants die down. Some trees have flaking or peeling bark, in warm shades of cinnamon, brown and russet; in others the bark is of a polished smoothness, and from rich mahogany to the purest, cleanest white. The smooth, coloured stems of dogwoods and other shrubs shine out in the sunlight, and the contorted hazel and the twisted willow can look superb, with writhing branches set against a clear blue sky.

The Tibetan cherry, *Prunus serrula*, with shining mahogany bark, rises from a cloud of *Carex comans* 'Bronze'. Behind, the parchment fronds of miscanthus contrast with shining evergreens.

TREE BARK

The Tibetan cherry, ***Prunus serrula***♡, is the perfect winter tree of an ideal size for small gardens, while larger gardens could accommodate a whole grove. It is grown not for its narrow willow-like leaves, its fleeting small white flowers in spring, nor its sparse production of cherries in the autumn, but for the glory of its bark, beautiful in all seasons, but especially eye-catching in winter when the low sun burnishes the polished stems, highlighting the chestnut-red surface. In its youth the trunk is a smooth shining red and almost irresistible to touch, but as the tree matures horizontal bands of corky brown develop, rather spoiling the smooth surface.

This lovely ornamental cherry is best planted to stand alone in grass or on a promontory in a border so that the trunk can be seen from all sides and can be stroked by all those who pass by. However, the roots are very shallow and can become a problem on grass paths. Occasional careful pruning can be undertaken to remove crossing branches and to keep the centre open so that the red stems can be admired. Surround the base with clumps of *Carex comans* 'Bronze', a froth of narrow grassy copper leaves echoing the colour of the cherry in a paler tone, and add groups of a very pale pink hellebore and a few snowdrops.

The Himalayan birch, ***Betula utilis***, is a wonderful tree for its beautiful bark in

THE ENDANGERED HIMALAYAN BIRCH

The name *utilis* means useful and in the Himalayas the trunks of these white birches do indeed serve many purposes, including forming makeshift bridges over rushing mountain torrents. They are also heavily used for firewood and as a result are becoming endangered in their native habitat.

OTHER TREES WITH ATTRACTIVE BARK *Acer* × *conspicuum* 'Phoenix' • *Acer* × *conspicuum* 'Silver Vein' •

The white peeling bark of the Himalayan birch *Betula utilis* var. *jacquemontii* contrasts with the ebony wands of *Cornus alba* 'Kesselringii'.

Betula utilis var. *jacquemontii*

shades from ivory white and rich cream to milky coffee and coppery brown. ***Betula utilis*** var. ***jacquemontii***♀, with white bark, is particularly delightful in winter, standing out against a backdrop of a dark hedge or forming a focal point in a lawn. It eventually makes a medium-sized tree with a delicate tracery of branches; the leaves are quite large for a birch but they do not cast a lot of shade. The foliage turns buttery yellow in autumn and then in late winter the leafless twigs are suddenly trembling with long pendulous yellow catkins. For a charming combination grow the autumn-flowering *Cyclamen hederifolium* var. *hederifolium* f. *albiflorum* at its feet, the white flowers followed by silver-marbled leaves, and to extend the season have yellow winter aconites nestling among the cyclamen leaves. These bulbous plants are quite happy drying out in the summer beneath the shade of the birch leaves.

Of the newer selections, ***Betula utilis* 'Silver Shadow'**♀ is a good choice, as is the Dutch-bred **'Doorenbos'**♀, both with gleaming white trunks. The latter is often offered as *Betula utilis* 'Snowqueen'; it displays white bark from the second or third year, earlier than most cultivars. ***Betula utilis* 'Grayswood Ghost'**♀ is a beautiful form with chalk-white bark; it looks wonderful rising from a bed of white winter-flowering heathers. ***Betula albosinensis*** var. ***septentrionalis***♀ has coppery-pink bark and looks particularly effective when grown in front of a beech hedge, its paler bark offset by the chestnut colour in the dying beech leaves. ***Betula albosinensis* 'Bowling Green'** has exquisite honey-coloured peeling bark that flutters in the breeze, looking like sheets of thin toffee in the winter sun.

CLEANING HIMALAYAN BIRCH

To keep the white trunk of Himalayan birch pristine all winter, scrub it with a stiff brush dipped in clean water. A light scrub will remove any green algae, which can spoil its good looks.

Betula albosinensis 'Bowling Green'

Arbutus menziesii • *Eucalyptus dalrympleana* • *Pinus sylvestris* • *Prunus maackii* • *Prunus serrulata* •

Betula ermanii

Acer grosseri var. hersii

Betula ermanii is perhaps the most tactile of the birches, with creamy-white, sometimes pinkish bark with pale brown corky horizontal markings; these are the lenticels (pores) through which the trunk of the tree breathes. The bark is smooth and silky, like fine kid leather.

All these birches can be planted in a group of three or more, where space permits; in smaller gardens, a young specimen can be cut back severely so that it shoots from the base to form a multi-

Acer griseum

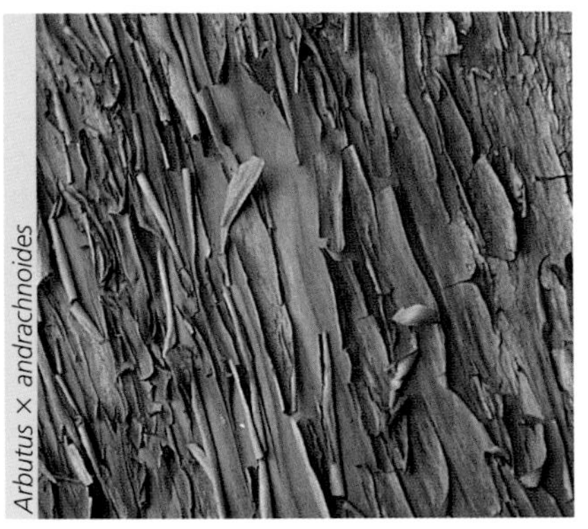
Arbutus × andrachnoides

stemmed tree, or three young plants could be placed in the same hole – triple the pleasure in one spot. All are best planted as young trees, when they will quickly become established. Some of the lower branches may be removed in early winter to expose more of the main stem and side branches. This also helps to thin the head of the tree, which will allow more rain and light to penetrate in spring and summer. The bark peels as the tree grows, revealing fresher, even whiter bark beneath; it may be tempting to strip the bark away, but it is better to let the wind gradually remove it. These birches are very hardy and will grow happily in almost any soil, but they can struggle when too dry in summer.

The snake-bark maples have delightful stems patterned like snakeskin. The Chinese ***Acer davidii* 'George Forrest'**♕ is a slight tree with green bark striped with silver, and ***Acer grosseri* var. *hersii***♕ has a zebra patterning of silver veins. The Japanese ***Acer rufinerve***♕ is a maple with a good shape and very decorative bark; the young shoots are bluish green, becoming striped in green and white as they mature, and the leaves turn vibrant red before they fall. ***Acer palmatum* 'Sango-kaku'**♕ (formerly 'Senkaki'), the coral-bark maple, is brighter still, with the young branches a brilliant coral red, and the paper-bark maple, ***Acer griseum***♕, when mature has bark that curls off in flakes, revealing the cinnamon underbark. These maples usually form only small trees and many can be cut back to produce numerous stems for greater impact. They are best grown in moist, neutral to acid soil, in an open situation.

Acer pensylvanicum♕, the North American snake-bark maple, is best represented by ***Acer pensylvanicum* 'Erythrocladum'**, which has new growth in a startling shrimp pink, maturing to orange-red with white stripes, and is particularly striking in winter. This is a rather delicate tree that needs careful positioning, perhaps in the shelter of other trees; it is most effective grown as a multi-stemmed specimen.

There are a few evergreen trees that are grown for the beauty of their bark: even though they do not shed their leaves in winter, their trunks can still be seen through the leaves. The handsome small tree ***Arbutus* × *andrachnoides***♕ has rich orange-red flaking bark, and the similar ***Luma apiculata***♕ has cinnamon bark that peels off, revealing the creamy underbark. The lacebark pine, ***Pinus bungeana***, with pale grey, flaking stems, was introduced from China many years ago but is seldom planted although easy to grow. It slowly forms a medium-sized tree, often branching from near the base, with bark that peels to produce a patchwork of grey, purple and green. ***Eucalyptus pauciflora* subsp. *niphophila***♕, the snow gum, has been collected from high up in the Snowy Mountains in Australia and is proving hardy in Britain; it grows slowly into a small tree with an attractive trunk dappled in green, grey and cream.

Eucalyptus pauciflora subsp. niphophila

STEMS

One of the most important contributions to the texture of the winter garden, and some of the brightest colours, come from the stems of the deciduous dogwoods (*Cornus*). If the plants are cut back frequently to encourage new growth, they produce yellow, orange, rich red or purple-black stems. Where space allows them to be planted en masse, the effect is dramatic, and even more so when they rise from a smooth carpet of snow.

The various cultivars of ***Cornus sericea*** (formerly *stolonifera*) and ***Cornus alba*** cover the full colour range. The best one for dark red stems is ***Cornus alba* 'Sibirica'**♀. Not only does ***Cornus alba* 'Elegantissima'**♀ have good red winter stems, its green leaves in summer are attractively margined with white. ***Cornus alba* 'Spaethii'**♀ has gold-edged leaves. ***Cornus alba* 'Kesselringii'**♀ has dark purple stems (see page 145) and needs placing against a light background in order to show up – perhaps near paving or in moist gravel. Several dogwoods have yellow stems: ***Cornus sericea* 'Budd's Yellow'** is worth trying for its rich yellow stems, while those of the popular ***Cornus sericea* 'Flaviramea'**♀ are yellow tinged with green, and ***Cornus sericea* 'White**

Cornus sericea 'Cardinal', framed by a clump of *Carex comans* 'Bronze' at the front and the dried reeds of *Miscanthus sinensis* behind.

PLANTING PARTNERS FOR WINTER STEMS

Cornus alba 'Sibirica'♀ with *Carex morrowii* 'Fisher's Form' (**1**); *Cornus sericea* 'Budd's Yellow' and *Rubus cockburnianus* (**2**); *Cornus sericea* 'Flaviramea'♀ with *Molinia caerulea* 'Variegata' (**3**); and *Cornus alba* 'Elegantissima'♀ with *Erica carnea* 'Springwood White'♀ (**4**).

Cornus sanguinea 'Midwinter Fire' and *Cornus sericea* 'Flaviramea' planted with *Phormium* 'Yellow Wave'.

Physocarpus opulifolius 'Dart's Gold'

Gold'♈ has bright yellow stems and summer leaves edged with cream. These are all strongly growing shrubs, layering to form thickets in the damp soil they prefer, and are not easy to include in a mixed border. They need to be planted in a wilder part of the garden, where they can be seen from a distance and where their coloured stems show up well if caught by the low winter sun. They will grow in shade if not too dry, but some of the colour effect will be lost. The variegated arum, *Arum italicum* subsp. *italicum* 'Marmoratum'♈, with its strikingly marbled leaves, makes a good winter companion, as do snowdrops, winter aconites and tellimas.

A fairly new arrival on the scene is ***Cornus sanguinea* 'Midwinter Fire'**. This is a drama queen of a plant: it starts its performance in late autumn, when the dying leaves take on tones of apricot and peach before dropping to reveal stems that are yellow in the centre of the bush, becoming orange and finally lobster red at the tips. This vigorous shrub can be hard pruned in the spring to induce new shoots, and a framework can be built up to form a rounded bush, which will aptly portray its name throughout the winter. It is a selection of the British native dogwood and can cope better with drier soil than some dogwoods do, but when established it is a rapidly suckering shrub so needs to be planted wisely. There is often confusion with the similar ***Cornus sanguinea* 'Winter Beauty'**, which is less vigorous and has darker green leaves in summer. ***Cornus sericea* 'Cardinal'** (see page 147) is a good substitute for either of these varieties of *Cornus sanguinea*. It has vigorous upright stems of orange-yellow, darker red near the tips.

A number of willows (***Salix***) also have good coloured stems in winter, among them the glowing yellow ***Salix alba* var. *vitellina***♈ and the brilliant orange-scarlet ***Salix alba* var. *vitellina* 'Britzensis'**♈ (see page 121). These can be pruned in the same way as dogwood to encourage them to produce bright new shoots.

***Physocarpus opulifolius* 'Dart's Gold'**♈ is a compact shrub with arching branches that, on a mature specimen, have fine peeling brown bark, shedding rather like a deer's antlers in autumn. It also has excellent summer foliage in soft gold, which does not burn in the sun. Although physocarpus are said to need a moist soil, this one seems to be able to grow almost anywhere.

PRUNING DOGWOOD FOR COLOURED STEMS

To make the most of dogwoods' coloured stems the plants need to be cut back in early spring. This is hard to do when the stems are still looking so attractive, but if left unpruned they produce lots of small twiggy growth and the colour will be lost. The best plan is to cut back half the stems to ground level each year, leaving the remainder to produce summer leaves, which in many forms are variegated in yellow or cream. These stems in turn can be cut back the following year. Controlled this way the plants will perform well throughout the year.

OTHER PLANTS FOR WINTER STEMS *Leycesteria formosa* • *Phyllostachys aureosulcata.* f. *aureocaulis* •

Some of the whitest stems in the winter garden come from ***Rubus cockburnianus***, often known as the ghost bramble, but you have to be brave to plant it – in fact, it has been referred to as 'vegetable barbed wire'. The cultivar ***Rubus cockburnianus* 'Goldenvale'**♀ has bright golden yellow fern-like foliage in summer and is slightly lower-growing than the species, its sharply prickled, arching stems often stretching to about 1.2m (4ft). The purple young stems are overlaid with a white coating that disappears with age, so they have to be regularly cut back (not a job to be relished). A really eye-catching grouping could be made by planting clumps of the black, grass-like *Ophiopogon planiscapus* 'Nigrescens'♀ beneath the rubus (see below).

Corylus avellana 'Contorta'

Growing *Corylus avellana* 'Contorta' in a pot restricts its size and displays its twisted stems and catkins to advantage.

In smaller gardens ***Rubus thibetanus***♀ would be a more suitable choice. Its form is more erect, with brown stems covered with a blue-white bloom, and it is less fiercely armed with thorns. The fern-like leaves are a delicate silver. It was introduced from China by Ernest Wilson in 1904 and is sometimes listed as *Rubus thibetanus* 'Silver Fern'.

***Corylus avellana* 'Contorta'**, the corkscrew hazel, is an interesting form with twisted, waving stems and golden yellow hanging catkins in late winter and early spring. It does well in any soil, in sun or shade, and eventually makes a large shrub although it can be slow to get going. Cut out any straight, upright stems at the base of the plant; these are usually suckers from the rootstock and quickly take over if allowed to. The summer foliage is puckered and distorted and not attractive; this is not a plant to feature at the front of a border.

WINTER IN SUMMER

A tangle of white arching stems in winter, *Rubus cockburnianus* 'Goldenvale'♀ is a mass of bright yellow leaves in summer. The black, grass-like perennial *Ophiopogon planiscapus* 'Nigrescens'♀ makes a good year-round partner.

Rubus phoenicolasius • *Salix alba* 'Dart's Snake' • *Salix daphnoides* • *Salix erythroflexuosum* • *Vaccinium corymbosum* •

Movement

Not all movement in the garden in winter is desirable, indeed the sight of trees bending in a gale and smaller plants swirling with the wind may cause concern rather than bring pleasure. However, some grasses and certainly bamboos bring an extra dimension with their swaying canes and rustling leaves, trees with pendulous branchlets catch the wind, and the stems of light evergreen shrubs wave in the breeze. Rippling, bubbling water and birds flitting from branch to branch also add life, as they do in any season.

The fine papery leaves and feathery plumes of miscanthus (above) and the delicate twigs of silver birch, *Betula pendula* (above right), are rarely still in the winter garden.

There is no more graceful tree than the silver birch, ***Betula pendula***. Tall, light and elegant, it is supple but strong-growing, eventually becoming a medium-sized tree, with height but little weight. The tracery of fine twigs seems to move effortlessly and gently, even in the strongest wind. ***Betula pendula*** **'Tristis'**♀ forms a graceful column with numerous slender, hanging branches that tremble in the wind. ***Populus tremula*** **'Pendula'**, the weeping aspen, is also a good choice, being particularly attractive in late winter when it is covered with long, dangling, purplish-grey catkins.

When grown naturally, **eucalyptus** is similar to birch in shape and character but the foliage is evergreen. Although the surfaces of the leaves are matt, their blue-green colour is very reflective, sparkling when stirred by the breeze in the winter sunlight. When grown as shrubs they are stooled (cut back to the ground every couple of years). This produces tall, straight shoots of juvenile foliage that is usually superior in colour to that of a mature plant allowed to grow naturally. These stems grow quickly and are pliable and supple. ***Eucalyptus gunnii***♀ is one of the best to grow in this way; it has round silver-blue juvenile leaves on tall, straight stems 2m (6ft) or more high.

Phillyrea angustifolia, a member of the olive family, is a medium-sized, light evergreen shrub with far more movement

BIRDS AND WATER

Birds bring movement into the garden, too: the solitary bird flying overhead, the excited flutter of a small flock flitting from tree to tree, and the busy activity of ground birds pecking for worms, lifting up the dead leaves and peering beneath. A bowl of water on the terrace will attract bathing birds that flutter and splash and chatter as they perform their ablutions.

BAMBOOS

Bamboos have a reputation for spreading uncontrollably but most of the clump-forming bamboos can easily be kept within bounds, and their evergreen leaves are welcome in winter, rustling as they catch the wind.

Fargesia murielae 'Simba'🏆 is a neat bamboo, reaching no more than 2m (6ft) high, that forms gracefully arching clumps of leafy canes with bright green leaves. One of the best bamboos for smaller gardens, it is also suitable for growing in a large tub.

Bamboos with rich yellow stems show up particularly well and, if there is space, the tall golden yellow *Phyllostachys vivax* f. *aureocaulis*🏆 is a good choice. A clump-forming bamboo that can reach spectacular heights of 6m (20ft) or so, it is best planted to stand alone, rising out of a sea of gravel or pebbles. Trimming back the lower leafy side branches to about 2m (6ft) shows the golden canes to advantage.

The black bamboo, *Phyllostachys nigra*🏆, reaching a mere 3.5m (12ft), has shining, arching, ebony stems, particularly effective when they catch the sun. This, too, is a clump former and can be divided when it gets too large. The young plants start green, attaining their full blackness in maturity.

Thamnocalamus crassinodus 'Kew Beauty' is an enchanting bamboo, with tiny leaves that flutter in the breeze. Growing up to 4m (13ft) high and arching over, the canes are white-bloomed when young, maturing to grey-blue.

Sasa veitchii (above and right) has purple canes and winter leaves in green striped with pale straw. A very attractive bamboo, it grows into a low, dense thicket, at its best in wilder areas of the garden.

All these are hardy evergreens that thrive in almost any conditions except a bog; all grow quickly to reach their ultimate height. To maintain a handsome specimen, remove some of the older stems from the centre, letting in air and light, and chop off any new shoots that appear beyond the clump.

than most other structure shrubs. It has arching stems weighed down with luxuriant masses of narrow, dark green foliage. ***Phillyrea latifolia*** is similar, with slightly broader leaves. In both species, the branches have the appearance of long sparkling rich green plumes that move in the wind.

Tall grasses such as ***Miscanthus*** and ***Calamagrostis*** look beautiful swaying in the wind but they can become a bit battered by late winter and may need to be cut down and the debris cleared away. The stems of miscanthus can be maintained for longer if the tops are removed when the leaves start to fall in midwinter. This will leave clumps of parchment reeds that still catch the light and provide structure in the winter garden.

Some of the smaller grasses are more resilient: the arching russet-brown stems and paler seedheads of the Japanese mountain grass, ***Hackonechloa macra***, lift with the passing currents of air, and the pale straw flowers of the lovely little ***Stipa tenuissima*** shimmer in the slightest breeze. (See also pages 30–31.)

MOVING WATER

Moving water brings life to a garden at any time of the year. In winter the sound of bubbling or splashing water has a cold crisp clarity that carries on the winter air as distinctly as the delicate scent of winter blooms. As winters in many areas become milder it is possible to keep water features running throughout the year. In colder regions it may be necessary to turn them off or drain them when freezing conditions threaten.

Taste

Picking herbs and vegetables straight from the garden is one of life's simple pleasures and, harvested in the peak of condition, they taste so much better than those bought at the supermarket. In many herbs the leaves are aromatic, appealing to our sense of smell as well as taste, and plants will give structure throughout the year, whether in pots by the house or in the garden proper.

Laurus nobilis

Parsley and sage are essential herbs for a sunny bed near the house.

Rosmarinus officinalis

HERBS FOR WINTER

Summer herbs can be spread around the garden, as a walk to collect fennel or chives is always a pleasure, but in winter herbs need to be grown in pots or beds near the house, since a dash through pouring rain or a howling gale to gather the stems is not so welcome.

Sweet bay (***Laurus nobilis***) is such a versatile evergreen. It can be grown as a large structure shrub or trimmed and trained into a variety of shapes. It is also ideal to grow in a pot near to the wall of the house in winter where it will enjoy good drainage and protection from freezing winds. The fragrant dark green leaves have been used for centuries to flavour both sweet and savoury dishes. The flavour is powerful so even a small plant will yield plenty for soups, stews and marinades. Underplant it with the golden form of marjoram (*Origanum vulgare* 'Aureum'🏆) and the prettily gold-variegated *Thymus* 'Doone Valley' (both described below) and you will have most of the ingredients for your own fresh bouquet garni in one container.

Parsley (***Petroselinum crispum***) can be grown in a large container by the kitchen door, in a sunny or lightly shaded spot, and it will be readily accessible for cutting throughout winter. Older plants will run out of steam eventually and will need to be replaced with vigorous young plants from time to time. It is not necessary to grow the parsley in isolation: in a large enough pot, it makes a pretty frilly base

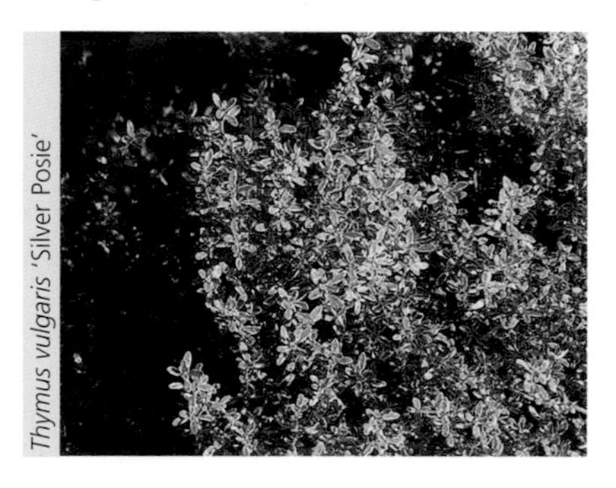
Thymus vulgaris 'Silver Posie'

Origanum vulgare 'Aureum'

Origanum vulgare 'Polyphant'

for a delicate summer-flowering clematis, such as *Clematis florida* var. *sieboldiana*.

Thyme (***Thymus vulgaris***) is a good winter herb and does well in a shallow terracotta pot in full sun; the plants will be kept neat by constant cutting. The plain green form is most commonly used for culinary purposes, but some of the variegated thymes are just as effective and look most attractive. ***Thymus vulgaris* 'Silver Posie'**, with silver-edged grey-green leaves, and ***Thymus* 'Doone Valley'**, with dark green leaves irregularly splashed with gold, are good varieties that stand up well to cold winter weather.

Rosemary (***Rosmarinus officinalis***), on the other hand, is better grown in a border; thriving in well-drained, sunny conditions, it copes easily with a dry spot near the house walls. ***Rosmarinus officinalis* 'Miss Jessopp's Upright'**♕ is the form most often seen, making a nice narrow column with aromatic leaves and bright blue flowers in early summer.

The purple sage, ***Salvia officinalis* 'Purpurascens'**♕, is an essential ingredient of sunny borders so there should always be plenty of plants to raid for the kitchen. The matt purple leaves take on an extra dimension when rimmed with frost and a trim in spring keeps the plants bushy. There are also coloured sages with golden leaves. ***Salvia officinalis* 'Icterina'**♕ is most commonly grown, with grey-green leaves edged and variegated with soft gold. The more tender ***Salvia officinalis* 'Tricolor'**, with leaves splashed with pink and ivory, is best grown in a pot.

Marjoram, or oregano, comes in different colours, varying from the plain green ***Origanum vulgare*** to the bright gold ***Origanum vulgare* 'Aureum'**♕ and the neatly cream-variegated form ***Origanum vulgare* 'Polyphant'**. Grow these marjorams in a sunny spot – in a bed or border or in containers; all will need clipping back to the ground in the autumn to encourage them to produce a flush of new leaves for the winter.

WINTER VEGETABLES

The early cottage gardeners depended on their gardens for winter vegetables but the choice was very poor. Winter cabbage was an early stalwart and there were always herbs such as thyme and sage, which were very necessary for flavouring food. There were the root vegetables to be dug and stored, and usually a bay tree – its leaves were used for flavouring and it was also a pleasant evergreen in winter.

Nowadays we have a much better selection if we wish to grow our own although, regrettably, most of us rely on the supermarkets bringing in produce from all round the world, with no regard to seasons.

In modern gardens, leeks (above) are a good standby, withstanding frost, and a row of leeks can be pulled all through the winter. The variety 'Toledo' is an old favourite, with very long stems. Another stalwart is purple-sprouting broccoli, a hardy brassica that can cope with wintry weather and still produce a succession of tasty flowerheads over several weeks in late winter. The variety 'Rudolf' is vigorous, but it needs to be grown in soil that is rich in organic material. Parsnips can be dug when the weather is mild, with a variety such as 'Gladiator' being lifted when it is still small and delicately flavoured.

There are also several varieties of cabbage that heart up in winter, with 'Tundra', 'Siberia' and 'Alaska' all bearing names denoting winter suitability. The savoy cabbages look decorative, too, with their dark blue-green, crinkled leaves with prominent veining, tinged with red in the variety 'January King'. The Brussels sprout called 'Revenge' matures for Christmas and continues cropping into late winter, and the newer 'Millennium' has an excellent flavour.

With increased interest in the ornamental qualities of vegetables as well as the taste, the kale 'Nero di Toscana Precoce' has become popular for winter harvest. Its long, puckered holly-green leaves look almost black in the garden and the flavour is strong and delicious, excellent with winter casseroles. The ornamental Swiss chard or leaf beet 'Bright Lights' adds colour to the vegetable garden or flower bed in early winter. The long succulent leaf-stalks are bright cherry red, glowing orange or fluorescent yellow, contrasting with the green spinach-like leaves.

There are lettuces that can be sown outside in winter and should produce fresh leaves for picking in the spring, although it would be advisable to cover them with a cloche; both 'Winter Density' and 'Wonder of Winter' would be suitable.

Rhubarb can be forced in specially designed pots for succulent early stems. For many years Blooms of Bressingham in Norfolk supplied a cultivar called 'Grandad's Favourite', selected by the late Alan Bloom for the best flavour and richest red colour of all the early rhubarbs. Unfortunately, it is now hard to find, but *Rheum × hybridum* 'Timperley Early'♕ and 'Victoria' are more readily available.

Author's choice:
favourite winter planting groups

Here is a selection of plants that have proved themselves to be reliable performers in the author's garden at White Windows in Longparish, Hampshire. All contribute colour and interest during the winter months. The treasured snowdrops, hellebores, aconites and cyclamen are to be admired for their brave display in the harshest of seasons. The evergreen shrubs and perennials are invaluable in all parts of the year, but especially in winter when their bright shining leaves provide much of the colour in the garden.

The Tibetan cherry, *Prunus serrula*, takes centre stage in the author's garden at White Windows in winter.

AROUND TREES WITH ATTRACTIVE BARK

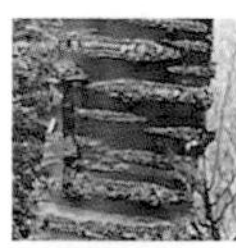

Prunus serrula♀ (page 144) Deciduous tree with willow-like leaves and white flowers in spring. Wonderful polished mahogany bark.

Carex comans 'Bronze' (page 31) Evergreen sedge with fine grass-like foliage of pale bronze, parchment at the tips. Dislikes very dry soils.

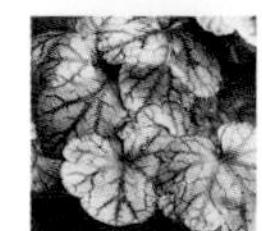

× *Heucherella* 'Quicksilver' (page 60) Evergreen perennial with dark wine leaves overlaid and etched with silver. Prefers semi-shade.

Helleborus × *hybridus* (white, cream or pink) (page 73) Semi-evergreen perennial with nodding flowers in late winter and spring.

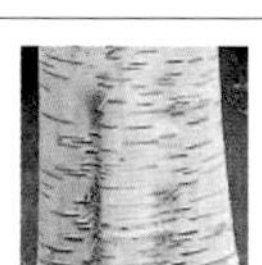

Betula utilis var. *jacquemontii*♀ (page 145) Deciduous tree with white bark throughout the year, more conspicuous in winter. Good as a focal point.

Eranthis hyemalis♀ (page 64) Corm producing bright yellow flowers from midwinter. Thrives in moist soil in semi-shade; dies down after flowering.

Arum italicum subsp. *italicum* 'Marmoratum'♀ (page 95) Patterned arrowhead leaves appear in winter and last until late spring. Spikes of red berries in autumn.

Galanthus nivalis♀ (page 71) Bulb producing green-marked white flowers in late winter. Prefers shade. For best results transplant after flowering.

FOR SHADE OR SEMI-SHADE

Osmanthus heterophyllus 'Variegatus'🏆 (page 49) Compact evergreen shrub. Small holly-like leaves, dark green edged white. Tiny fragrant flowers in autumn.

Skimmia japonica 'Rubella'🏆 (page 49) Evergreen shrub with red buds in winter opening to fragrant creamy-pink flowers in spring.

Tellima grandiflora Rubra Group (page 102) Evergreen perennial with green foliage turning red in winter. Small greenish-white flowers in early summer.

Bergenia 'Bressingham Ruby' (page 54) Evergreen perennial with shining green leaves, dark mahogany in winter. Rich purple-red flowers in early spring.

Epimedium × *rubrum*🏆 (page 55) Evergreen perennial. Copper new leaves; mature foliage copper-red in winter. Tiny red and yellow flowers in spring.

Galanthus 'S. Arnott'🏆 (page 71) Bulb producing relatively large, pleasantly scented white flowers on tall stems in late winter. Fine foliage.

FOR SHADE OR SEMI-SHADE AT THE BASE OF A WALL OR FENCE

Hedera colchica 'Sulphur Heart'🏆 (page 99) Evergreen climber. Large, glossy, dark green leaves marked with bright green and gold. Also good as ground cover.

Sarcococca hookeriana var. *digyna*🏆 (page 143) Evergreen shrub with narrow leaves on dark upright stems. Tiny pinkish flowers in winter. Very fragrant.

Euonymus fortunei 'Emerald 'n' Gold'🏆 (page 98) Evergreen shrub with small, green and gold leaves on lax spreading stems. Easy to grow.

Asplenium scolopendrium🏆 (page 91) Evergreen fern with long, tongue-shaped, shining bright green leaves in a shuttlecock rosette.

Narcissus 'Tête-à-tête'🏆 (page 66) Bulb producing short stout stems, each with two or three yellow flowers in late winter. Easy and reliable.

Ajuga reptans 'Atropurpurea' (page 103) Evergreen perennial with waved and puckered, shining dark purple leaves. Spikes of sapphire flowers in spring.

WINTER FOLIAGE FOR SUN

Pittosporum tenuifolium 'Irene Paterson'🏆 (page 92) Evergreen shrub with small, waved, pale green leaves mottled creamy white, on dark stems. Slow-growing.

Brachyglottis 'Sunshine'🏆 (page 105) Evergreen shrub with grey-green felted leaves on grey stems. Hardy. Cut back in summer for best winter foliage.

Euphorbia BLACKBIRD ('Nothowlee') (page 130) Evergreen perennial with red-black leaves on sturdy upright stems. Lime-green flowers in early spring.

Stipa tenuissima (page 30) Evergreen grass with fine hair like leaves of soft green and golden buff. Seeds freely on light soils.

Ophiopogon planiscapus 'Nigrescens'🏆 (page 103) Evergreen perennial with striking black strap-shaped leaves in clumps from spreading rhizomes.

Cyclamen coum🏆 (page 65) Corm with pretty silver and green marbled leaves. Delicate flowers in white to reddish pink in midwinter. Dies down after flowering.

WINTER FLOWERS AND FOLIAGE FOR SUN

Euphorbia characias SILVER SWAN ('Wilcott') (page 93) Evergreen, upright perennial with white-striped green leaves and white and green flowers in spring.

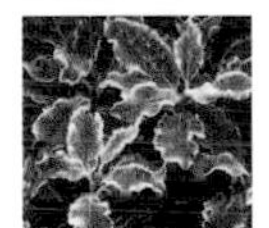
Pittosporum tenuifolium 'Tom Thumb'🏆 (page 102) Dwarf, compact evergreen shrub with waved, shining green leaves turning dark wine-purple.

Helleborus × *hybridus* (white or cream) (page 73) Clump-forming semi-evergreen perennial with nodding flowers in late winter and early spring.

Uncinia rubra (page 31) Evergreen sedge forming low mounds of shiny spiky foliage turning rich reddish bronze in winter.

Bergenia 'Eric Smith' (page 54) Evergreen perennial with large, leathery, shining leaves turning rich bronze-red in winter. Dark pink flowers in early spring.

Heuchera 'Plum Pudding' (page 59) Evergreen perennial. Wine-purple leaves lightly overlaid with silver on long stalks. Small cream flowers in summer.

Index

Page numbers with suffix 'b' indicate plants listed across bottom of page. Suffix 'i' refers to illustrations.

A

Abelia
A. 'Edward Goucher' 24
A. × *grandiflora* 24i
A. × *g.* 'Francis Mason' 24
Abeliophyllum distichum 136i
Abies
A. koreana 'Silberlocke' 50b
A. nordmanniana 'Golden Spreader' 50b, 96b
Acacia dealbata 82i, 142b
Acanthus spinosus 119
Acer
A. × *conspicuum*
A. × *c.* 'Phoenix' 144b
A. × *c.* 'Silver Vein' 144b
A. davidii 'George Forrest' 146
A. griseum 146i
A. grosseri var. *hersii* 146i
A. palmatum 115
A.p. 'Sango-kaku' (*A.p.* 'Senkaki') 146
A. pensylvanicum 'Erythrocladum' 146
A. rufinerve 146
acid soil 112–15
Aconitum 'Stainless Steel' 36
Acorus gramineus 'Ogon' 30b, 97
Adiantum venustum 103
Agave americana 52b
AGM (Award of Garden Merit) 5
Ajuga
A. pyramidalis 'Metallica Crispa' 103
A. reptans
A.r. 'Atropurpurea' 103i, 129i, 155i
A.r. 'Multicolor' 95
Alnus
A. incana 'Aurea' 120i
A. × *spaethii* 120b
alpine mint bush see *Prostanthera cuneata*
Amelanchier lamarckii 32
Andromeda polifolia 115, 120b
Anemanthele lessoniana 30i, 100b, 129
Anemone
A. blanda 81i, 113i
A.b. 'White Splendour' 39, 54
A. nemorosa
A.n. 'Buckland' 35
A.n. 'Vestal' 53
Anglesey Abbey, Cambridgeshire 10
Anthemis tinctoria 'Sauce Hollandaise' 35
Arabis
A. alpina subsp. *caucasica* 31
A. procurrens 'Variegata' 60i–61
Arbutus
A. × *andrachnoides* 81i, 146i
A. menziesii 145b
A. unedo 81
architectural plants 52–3
Arisaema candidissimum 38
Arum italicum subsp. *italicum* 'Marmoratum' 27i, 41, 95i, 154i
Arundo donax var. *versicolor* 30b
Asarum
A. caudatum 123
A. europaeum 56i, 74, 90, 123i
Asperula aristata subsp. *scabra* 76
Asplenium scolopendrium 74, 91i, 155i
A.s. Cristatum Group 91
A.s. 'Kaye's Lacerated' 91, 130
Astelia 53
A. chathamica 53i, 94i, 106i, 131i
A. nervosa 53
A.n. 'Westland' 52b
Aster
A. ericoides 29
A. lateriflorus 29
Astilbe 'Inshriach Pink' 38
Astrantia 'Hadspen Blood' 37
Athyrium niponicum 38
Atriplex halimus 105, 118i
Aucuba 99, 116
A. japonica 48b
A.j. 'Crotonifolia' 116i
A.j. 'Golden King' 99, 116
A.j. 'Marmorata' 99
A.j. 'Pepperpot' 99i
A.j. 'Variegata' 116
autumn 24–31
Award of Garden Merit (AGM) 5
Azara microphylla 136
A.m. 'Variegata' 124b

B

bamboos 151
bark, textured/patterned 144–6
Belfast, Botanic Gardens 8i
Berberis darwinii 88b
Bergenia 13i, 54–5,102
B. 'Baby Doll' 39
B. 'Bressingham Ruby' 54i, 155i
B. cordifolia 'Tubby Andrews' 95i
B. 'Eric Smith' 54i, 155i
B. purpurascens 110b
B. × *schmidtii* 54
B. stracheyi 54
B. 'Wintermärchen' (WINTER FAIRY TALES) 54
die-back in 41
Beschorneria yuccoides 52b
Betula
B. albosinensis
B.a. 'Bowling Green' 145i
B.a. var. *septentrionalis* 145
B. ermanii 146i
B. nigra 100, 120i
B. pendula 119i, 150i
B.p. 'Tristis' 119, 150
B. utilis 144–5
B.u. 'Doorenbos' 145
B.u. 'Grayswood Ghost' 145
B.u. var. *jacquemontii* 12, 34i–35i, 103, 145i, 154i
B.u. 'Silver Shadow' 145
B.u. 'Snowqueen' 145
cleaning bark of 145
birch see *Betula*
birds 28
Bloom, Adrian 9
Bloom, Alan 153
blue rue see *Ruta graveolens*
borders/beds
clearing/mulching 17
design of 18
botanical names 20–1
botanic gardens 8, 10, 11
Bowles, E.A. 68
Bowles' golden grass see *Milium effusum* 'Aureum'
box see *Buxus*
Brachyglottis
B. 'Sunshine' 35, 105i, 111, 155i
B. 'Walberton's Silver Dormouse' 105i
Brunnera 33, 117
B. macrophylla
B.m. 'Jack Frost' 33
B.m. 'Looking Glass' 33
Buddleja
B. 'Lochinch' 36
pruning 19i
bulbs 41, 64–71
bringing indoors 14
under rhododendrons 113
butcher's broom see *Ruscus aculeatus*
Butia capitata 53b
Buxus
box blight 45
B. microphylla
B.m. 'Faulkner' 45
B.m. 'John Baldwin' 45
B. sempervirens 45i, 89, 116b
B.s. 'Argenteovariegata' 39
B.s. 'Elegantissima' 45i, 131i
B.s. 'Suffruticosa' 45i, 89i
B. sinica 46
B.s. var. *insularis*
B.s.v.i. 'Justin Brouwers' 46
B.s.v.i. 'Winter Gem' 46

C

Calamagrostis 30, 151
C. × *acutiflora* 'Karl Foerster' 30
C. brachytricha 30i, 39
Callicarpa bodinieri 27
C.b. var. *giraldii* 'Profusion' 27i
Calluna vulgaris 115
C.v. 'Ariadne' 100i
C.v. 'Beoley Gold' 115
C.v. 'Joy Vanstone' 100
C.v. 'Robert Chapman' 115
C.v. 'Silver Fox' 115
C.v. 'Silver Queen' 106b
C.v. 'Wickwar Flame' 100, 115
Cambridge University Botanic Garden 10
Camellia 112–13
C. 'Cornish Snow' 112i, 113
C. 'Cornish Spring' 114b
C. japonica 112, 124b
C.j. 'Apple Blossom' 113
C.j. 'Nobilissima' 8i, 112–13
C. sasanqua 112, 142b
C.s. 'Crimson King' 112i
C.s. 'Narumigata' 112
C. × *vernalis* 113
C. × *v.* 'Yuletide' 113
C. × *williamsii*
C. × *w.* 'Jury's Yellow' 113
C. × *w.* 'Saint Ewe' 112i, 113
introduction of 8
Campanula persicifolia 36
Cardamine quinquefolia 33i
cardoon see *Cynara cardunculus*
Carex
C. buchananii 31, 74, 100, 101i, 131
C. comans 74, 83, 110b
C.c. 'Bronze' 31i, 39, 100i, 144i, 147i, 154i
C.c. 'Frosted Curls' 31
C.c. 'Taranaki' 100
C. conica 'Snowline' 31, 94, 95i
C. dipsacea 31, 101
C. morrowii
C.m. 'Fisher's Form' 147i
C.m. 'Variegata' 31b, 94
C. oshimensis 'Evergold' 31, 98b
C. testacea 101
pruning 101
Carpinus betulus 100i
Caryopteris × *clandonensis* 'Worcester Gold' 39
Ceanothus
C. arboreus 'Trewithen Blue' 126b
C. thyrsiflorus 'Skylark' 88b
Ceratostigma willmottianum 36
Chaenomeles 126
C. speciosa
C.s. 'Geisha Girl' 126
C.s. 'Moerloosei' 39, 126
C.s. 'Nivalis' 126i
C. × *superba*
C. × *s.* 'Crimson and Gold' 126
C. × *s.* 'Pink Lady' 126i
chalky soil 116–17
Chamaecyparis
C. lawsoniana
C.l. 'Lutea' 97
C.l. 'Stardust' 97
C.l. 'Winston Churchill' 97i
C. obtusa 'Nana Gracilis' 90i
C. pisifera 'Boulevard' 51b
C. thyoides 'Ericoides' 51b
Chamaerops humilis 52
C.h. var. *argentea* 52, 131i
Chasmanthium latifolium 31, 123
cherry laurel see *Prunus lusitanicus*
Chiastophyllum oppositifolium 'Jim's Pride' 94i
Chimonanthus praecox 84b, 136
C.p. 'Grandiflorus' 136–7
C.p. 'Luteus' 137i
Chionodoxa
C. luciliae 69i, 81i, 113, 128b
C. sardensis 68b
Choisya
C. GOLDFINGERS ('Limo') 96i
C. ternata 35, 88i, 125, 139i
C.t. SUNDANCE ('Lich') 96i
Christmas box see *Sarcococca*
Christmas rose see *Helleborus niger*
Chusquea culeou 120b
Cistus
C. × *cyprius* var. *ellipticus* 'Elma' 88
C. × *hybridus* 89b
clay soil 110–11
Clematis
C. 'Alba Luxurians' 35
C. 'Arabella' 39
C. armandii 127
C. cirrhosa
C.c. var. *balearica* 15, 116b, 127i
C.c. var. *purpurascens* 'Freckles' 127i
C.c. 'Wisley Cream' 127i
C. florida var. *sieboldiana* 153
C. 'Little Nell' 110
C. napaulensis 127
C. ROSEMOOR ('Evipo002') 126
C. viticella 35
climate change 40–1
climbers 19, 127
compost 17
conifers 50–1, 90, 97, 101
containers/pots 128–31
Convolvulus cneorum 106b
Cordyline 52–3
C. australis 52i
C.a. 'Red Sensation' 52
C.a. 'Torbay Dazzler' 53i
C.a. 'Torbay Red' 53i

corkscrew hazel see *Corylus avellana* 'Contorta'
Cornus 34, 147–8
C. alba 121, 147
C.a. 'Elegantissima' 118b, 147i
C.a. 'Kesselringii' 103i, 145i, 147
C.a. 'Siberian Pearls' 26b
C.a. 'Sibirica' 105, 121b, 131, 147i
C.a. 'Sibirica Variegata' 34–5
C.a. 'Spaethii' 147
C. mas 81i–82, 84i
C.m. 'Aurea' 82
C.m. 'Variegata' 35, 82
C. sanguinea
C.s. 'Midwinter Fire' 34–5, 148i
C.s. 'Winter Beauty' 148
C. sericea (*C. stolonifera*) 121, 147–8
C.s. 'Budd's Yellow' 121b, 147i
C.s. 'Cardinal' 147i, 148
C.s. 'Flaviramea' 147i, 148i
C.s. 'White Gold' 147–8
pruning 19, 148
Corokia × *virgata* 'Frosted Chocolate' 100b
Coronilla valentina subsp. *glauca*
C.v.s.g. 'Citrina' 126i
C.v.s.g. 'Pygmaea' 126
C.v.s.g. 'Variegata' 126
Cortaderia selloana
C.s. 'Aureolineata' 99b, 141i
C.s. 'Pumila' 31b
Corydalis
C. malkensis 68b
C. solida 38
Corylus avellana 'Contorta' 15i, 149i
Cotoneaster 26, 28i
C. atropurpureus 'Variegatus' 124
C. dammeri 26
C. franchetii 26, 110i, 116b
C. frigidus 'Cornubia' 26
C. horizontalis 124i
C. 'Hybridus Pendulus' 26
C. lacteus 26i
C. × *suecicus* 'Coral Beauty' 26
cotton lavender see *Santolina*
Crocus 67
C. chrysanthus
C.c. 'Gipsy Girl' 68i
C.c. hybrids 68
C.c. 'Ladykiller' 68, 128b
C.c. 'Snow Bunting' 68
C. etruscus subsp. *flavus* 69b
C. goulimyi 69
C. sieberi 67
C.s. 'Albus' (*C.s.* 'Bowles' White') 67i
C.s. subsp. *sublimis* 'Tricolor' 68
C. tommasinianus 62–3i, 67i
C.t. 'Ruby Giant' 68i
C.t. 'Whitewell Purple' 67i–68
Cryptomeria 50–1
C. japonica
C.j. 'Elegans Compacta' 50
C.j. Elegans Group 50i, 102
C.j. 'Elegans Nana' 50–1, 102
C.j. 'Pygmaea' 51
C.j. 'Vilmoriniana' 102
Crystal Palace, London 8
cultivar, definition of 21
Cupressus arizonica var. *glabra* 'Blue Ice' 51i, 107i
curry plant see *Helichrysum italicum*
cut flowers 14–15
Cyclamen 64–5
C. coum 37, 38, 39, 65i, 103, 113i, 155i
C.c. Pewter Group 65
C.c. Silver Group 65
C. hederifolium 65i, 123i
C.h. 'Silver Cloud' 106i
C.h. var. *hederiflorum* f. *albiflorum* 145
C. Miracle series 128b
Cynara cardunculus 104i, 105–6

D

Danae racemosa 47i
Daphne 117, 137
D. bholua 84b, 137i
D.b. 'Darjeeling' 137
D.b. var. *glacialis* 'Gurkha' 137
D.b. 'Jacqueline Postill' 126b, 137i
D.b. 'Peter Smithers' 137
D. jezoensis 137i
D. laureola 117, 122i–123
D. mezereum 117i
D.m. f. *alba* 117
D. odora 137
D.o. 'Aureomarginata' 125, 137i
Daphniphyllum himalaense subsp. *macropodum* 116i
daylily see *Hemerocallis*
Deschampsia 121
D. cespitosa 31
D.c. 'Bronzeschleier' (BRONZE VEIL) 31, 121i
D.c. 'Goldtau' (GOLDEN DEW) 31, 121
Dianthus 56i, 117i
D. 'Mrs Sinkins' 37
Dicentra
D. 'King of Hearts' 38
D. spectabilis 41
D. 'Stuart Boothman' 39
Digitalis purpurea 73, 122i
dogwood see *Cornus*
Drimys winteri 114b, 126i
Dryopteris erythrosora 74, 91i, 130

E

Echinacea purpurea 37
Edgeworthia 138
E. chrysantha 84b, 138i
E.c. 'Red Dragon' 138i
Edinburgh, Royal Botanic Garden 11
Elaeagnus 98–9, 138
E. × *ebbingei* 48b, 138
E. × *e.* 'Gilt Edge' 99
E. × *e.* 'Limelight' 99i
E. pungens 138
E.p. 'Frederici' 99i
E.p. 'Goldrim' 99
E.p. 'Maculata' 98–9i
Epimedium 37, 55–6, 100, 122
E. alpinum 55
E. davidii 100, 122
E. grandiflorum 56
E. × *perralchicum* 56
E. × *p.* 'Wisley' 56
E. × *rubrum* 55i–56, 100, 122, 123i, 155i
E. × versicolor
E. × *v.* 'Neosulphureum' 55i, 56
E. × *v.* 'Sulphureum' 56
Eranthis hyemalis 64i, 81i, 108–9i, 154i
E.h. 'Guinea Gold' 38, 64i
Erica
E. arborea 83i
E.a. 'Albert's Gold' 83
E.a. 'Estrella Gold' 83
E. carnea 83, 115
E.c. 'Ann Sparkes' 83, 96
E.c. 'December Red' 83
E.c. 'Foxhollow' 96i
E.c. 'Myretoun Ruby' 60i, 83i, 115i
E.c. 'Springwood White' 83i, 115, 129b, 147i
E.c. 'Vivellii' 83
E.c. 'Winter Beauty' 83, 136
E. × *darleyensis* 83, 115, 118b, 142b
E. × *d.* 'Arthur Johnson' 83i
E. × *d.* 'Kramer's Rote' 83
E. × *d.* 'White Perfection' 83i
E. erigena 83
E.e. 'Golden Lady' 83
E.e. 'Irish Dusk' 83
E. vagans 'Valerie Proudley' 96b
Eriobotrya japonica 124b
Eryngium variifolium 29, 95, 118
Escallonia laevis 'Gold Brian' 97b
Eucalyptus 150
E. dalrympleana 145b
E. gunnii 150
E. pauciflora subsp. *niphophila* 146i
E. perriniana 107b
Euonymus 116–17
E. fortunei 93, 116
E.f. BLONDY ('Interbolwi') 98
E.f. 'Emerald Gaiety' 92i, 93, 94i, 116
E.f. 'Emerald 'n' Gold' 34i, 35i, 74i, 98i, 116–17, 127b, 155i
E.f. 'Silver Queen' 47i, 117i, 125b
E. japonicus
E.j. 'Chollipo' 84i, 117i
E.j. 'Latifolius Albomarginatus' 117
Eupatorium ligustrinum 111
Euphorbia 56–7, 103, 107
E. amygdaloides 57
E.a. 'Purpurea' 57i
E.a. var. *robbiae* 90i
E. BLACKBIRD ('Nothowlee') 57, 103, 130i, 155i
E. 'Blue Haze' 57, 107
E. characias 56, 57i, 94i
E.c. SILVER SWAN ('Wilcott') 57i, 93i, 129, 155i
E.c. subsp. *wulfenii* 56–7, 107i
E.c.s.w. 'Lambrook Gold' 57
E. griffithii 32–3
E. × *martini* 57i, 60
E. myrsinites 57i, 107
E. nicaeensis 107
E. schillingii 33i
handling 57
evergreens 42–61
big-leaved 88–9
classic varieties 44–9
conifers 50–1, 90, 97, 101
low-growing 90
for shady walls 124–5
small-leaved 89
for sunny walls 126–7

F

Fargesia murielae 'Simba' 151
Farrer, Reginald 143
Fascicularia bicolor 53b
× *Fatshedera lizei* 12, 38, 124–5i
Fatsia japonica 88, 89i, 124
F.j. 'Variegata' 92b
fences/walls, plants for 124–7
ferns 91
Festuca
F. glauca 'Elijah Blue' 107i
F. valesiaca 'Silbersee' (SILVER SEA) 1–7
flowers
cut 14–15
winter 62–85
foliage 86–107
brown/tan/orange 100–1
gold 96–9
green 88–91
multicoloured 95
red/purple/black 102–3
scented 139
silver/blue 104–7
variegated 92–5
formal gardens 9, 13
Forsythia ovata 'Tetragold' 15
Fortune, Robert 82
fragrant plants 136–43
Fraxinus excelsior 'Jaspidea' 121b
frost, protection from 18
fruit, ornamental 28
Fuchsia 'Tom Thumb' 37, 38

G

Galactites tomentosa 95i
Galanthus 20i–21i, 70i–71i, 81i, 108–9i
G. 'Atkinsii' 71i
G. caucasicus 21
G. 'Desdemona' 71
G. 'Dionysus' 70b
G. elwesii 71i, 107
G. ikariae 20, 38, 71i, 90i
G. 'John Gray' 70b
G. 'Magnet' 71i
G. nivalis 11, 20–1, 70–1i, 122b, 123i, 129i, 154i
G.n. f. *pleniflorus* 'Flore Pleno' 71i
G.n. Scharlockii Group 70b
G.n. 'Tiny' 38
G.n. 'Viridapice' 71b
G. plicatus 71i
G.p. 'Augustus' 71b
G.p. 'Trym' 71
G.p. 'Wendy's Gold' 71
G. reginae-olgae 20, 69
G. 'S. Arnott' 71i, 76i, 125, 155i
G. 'Straffan' 71b
G. woronowii 71b
where to see 10, 11i
Galium aristatum 35
gardens to visit 10–11
Garrya
G. elliptica 84, 125
G.e. 'James Roof' 84i, 125
G. × *issaquahensis* 'Glasnevin Wine' 125i
Gaultheria 27, 115
G. mucronata
G.m. 'Bell's Seedling' 115i
G.m. 'Mulberry Wine' 26b
G.m. 'Wintertime' 115
G. procumbens 115, 130b
Geranium
G. macrorrhizum 55
G.m. 'Album' 125
G. 'Mavis Simpson' 37
G. nodosum 55
G. × *oxonianum*
G. × *o.* 'Spring Fling' 37

G. × o. 'Walter's Gift' 37
G. phaeum 48
G. pratense 'Mrs Kendall Clark' 36
G. pyrenaicum 'Isparta' 36
G. 'Sue Crûg' 38
G. wallichianum 'Buxton's Variety' 70
golden feverfew see Tanacetum parthenium 'Aureum'
grasses/sedges 30–1
Griselinia littoralis 48b, 88–9i
G.l. 'Dixon's Cream' 93b
ground-cover conifers 51, 101

H
Hacquetia epipactis 38, 79i
H.e. 'Thor' 79i
Hakonechloa macra 31, 151
H.m. 'Aureola' 101i
Ham House, Surrey 11
Hamamelis 136i, 140–1
H. 'Brevipetala' 114b, 140i
H. × intermedia
H. × i. 'Aphrodite' 138, 141i
H. × i. 'Arnold Promise' 140
H. × i. 'Diane' 140i
H. × i. 'Harry' 140–1i
H. × i. 'Jelena' 140i, 141i
H. × i. 'Orange Peel' 141i
H. × i. 'Pallida' 85b, 140i, 141i
H. × i. 'Primavera' 140
H. × i. 'Ruby Glow' 141
H. mollis 140i
H. vernalis 141i
H.v. 'Sandra' 141
H. virginiana 141
National Collection 10, 141
Harlow Carr, RHS Garden, Yorkshire 11
heathers 83, 96, 115
Hebe 89, 96
H. albicans 103, 105i
H. 'Caledonia' 103
H. 'Emerald Gem' 89, 130b
H. × franciscana 'Variegata' 93b
H. macrantha 89
H. ochracea 'James Stirling' 96i
H. pimeloides 'Quicksilver' 107b
H. pinguifolia 'Pagei' 105
H. rakaiensis 31, 89i
H. recurva 'Boughton Silver' 105
H. 'Red Edge' 84i, 103i, 117b
H. topiaria 89
Hedera 46–7, 88, 93, 99, 124
H. canariensis 'Gloire de Marengo' 93i, 125b
H. colchica 99
H.c. 'Dentata Variegata' 99i
H.c. 'Sulphur Heart' ('Paddy's Pride') 82i, 99i, 125i, 155i
H. helix 46–7
H.h. 'Amberwaves' 48, 97
H.h. 'Duckfoot' 47
H.h. 'Erecta' 47
H.h. 'Glacier' 93
H.h. 'Goldchild' 98i, 99
H.h. 'Goldheart' ('Oro di Bogliasco') 99
H.h. 'Green Ripple' 47i
H.h. 'Manda's Crested' 90
H.h. 'Pedata' 47
H.h. f. poetarum 'Poetica Arborea' 46i, 47
H.h. 'Spetchley' 47
symbolism of 46
Helianthemum 90
H. 'The Bride' 105i
Helianthus 'Lemon Queen' 35
Helichrysum italicum 104, 118b
H.i. 'Korma' 104i
Helleborus 36i, 41, 72i–77i
cutting flowers of 14, 74
H. argutifolius 42–3i, 55i, 74i–76
H. 'Briar Rose' 75i
H. corsicus see H. argutifolius
H. dumetorum 76i
H. × ericsmithii 77i
H. foetidus 55, 76, 90, 113i, 123, 131
H.f. Wester Flisk Group 76i, 123i
H. × hybridus 13, 36, 38, 72i–73i, 122, 131, 154i, 155i
H. × h. 'Usha' 123i
H. lividus 77
H. multifidus subsp. hercegovinus 36
H. niger 13, 21i, 38, 39, 72, 75i, 129i
H.n. Blackthorn Group 75
H.n. 'White Magic' 75
H. × nigercors 75
H. odorus 72
H. orientalis 72, 134i
H.o. Early Purple Group 66
H. 'Pink Ice' 75
H. purpurascens 72
H. × sternii 77, 131
H. × s. Blackthorn Group 76i, 77
H. × s. 'Boughton Beauty' 77
H. thibetanus 75i
H. torquatus 72
H. vesicarius 75
planting partners for 74
Hemerocallis 36i, 72i
herbaceous perennials 17–18, 32–3
herbs 152–3
Heuchera 37, 58–9, 102–3
H. 'Amber Waves' 31
H. americana 'Ring of Fire' 59
H. 'Beauty Colour' 58i, 102
H. 'Can-can' 130
H. 'Caramel' 101b
H. 'Cascade Dawn' 130
H. 'Chocolate Ruffles' 59, 103
H. CRÈME BRÛLÉ ('Tnheu041') 31, 59i, 95i, 131
H. 'David' 59i
H. EBONY AND IVORY ('E and I') 102
H. 'Green Spice' 59i
H. KEY LIME PIE ('Tnheu41') 59i, 130
H. LICORICE ('Tnheu044') 130
H. 'Lime Rickey' 59
H. 'Marmalade' 31, 37i, 59
H. micrantha var. diversifolia 'Palace Purple' 58
H. 'Obsidian' 102
H. 'Plum Pudding' 59i, 129, 155i
H. 'Prince' 130
H. 'Purple Petticoats' 37
H. 'Stormy Seas' 58i, 102
H. 'Venus' 59
× Heucherella 60
× H. 'Quicksilver' 60i, 154i
Hippophae rhamnoides 27b
holly see Ilex
honesty see Lunaria annua
hornbeam see Carpinus betulus
Hosta 'Hydon Sunset' 35
hyacinth 69i
hybrid, definition of 21
Hydrangea
H. paniculata 24i
H. serratifolia 127b

I
Iberis sempervirens 90
Ilex 44–5, 88, 98
I. × altaclerensis
I. × a. 'Camelliifolia' 88i
I. × a. 'Golden King' 45, 98i
I. × a. 'Lawsoniana' 98
I. aquifolium
I.a. 'Argentea Marginata' 45, 94i
I.a. 'Bacciflava' 44i–45
I.a. 'Ferox' 44i, 45, 88, 117b
I.a. 'Ferox Argentea' 93i
I.a. 'Flavescens' 97b
I.a. 'Handsworth New Silver' 36, 45, 93
I.a. 'J.C. van Tol' 26i, 45
I.a. 'Madame Briot' 44i–45
I.a. 'Myrtifolia Aurea Maculata' 45
I.a. 'Pyramidalis' 44i–45
I. cornuta 131i
I. crenata 45
I.c. 'Convexa' 44i–45
I.c. 'Golden Gem' 44i–45, 96i
I. × meserveae 106–7
I. × m. BLUE PRINCE 107i
I. × m. BLUE PRINCESS 107
symbolism of 46
indoor pots/flowers 14–15
Ipheion
I. 'Alberto Costillo' 68, 69i
I. uniflorum 68, 90
I.u. 'Charlotte Bishop' 68
I.u. 'Wisley Blue' 68, 69i
Iris
I. danfordiae 129b
I. foetidissima 27i, 61, 117i, 122b, 139
I.f. 'Variegata' 61i, 94
I. 'George' 67, 129b
I. 'Harmony' 67i
I. histrioides 66
I. 'Joyce' 67
I. 'J.S. Dijt' 69b
I. 'Katharine Hodgkin' 66
I. lazica 77
I. 'Pauline' 67
I. reticulata 66, 67i
I. unguicularis 15, 77, 78i, 117i, 119b
I.u. 'Alba' 77
I.u. 'Mary Barnard' 77i
I.u. 'Walter Butt' 77i
Itea
I. ilicifolia 125b
I. virginica 'Henry's Garnet' 115i
ivy see Hedera

J
Jasminum nudiflorum 13, 48, 82i–84, 124i
Jerusalem sage see Phlomis fruticosa
Juncus patens 'Carman's Gray' 121
Juniperus
J. chinensis 'Pyramidalis' 106
J. communis
J.c. var. depressa 101
J.c. 'Repanda' 101
J. horizontalis
J.h. 'Bar Harbor' 51
J.h. 'Wiltonii' 51
J. × pfitzeriana 'Old Gold' 51i
J. scopulorum
J.s. 'Blue Arrow' 106
J.s. 'Skyrocket' 106
J. squamata
J.s. 'Blue Carpet' 106i
J.s. 'Blue Star' 106i, 119b
J. virginiana 106
J.v. 'Burkii' 106

K
Kew, Royal Botanic Gardens 11, 58

L
labels 18
Lamium maculatum 61
L.m. 'White Nancy' 61i, 94i
Lathyrus vernus 91
latin names 20–1
Laurus nobilis 152i
L.n. 'Aurea' 97i
Lavandula 104, 139i
L. angustifolia 104, 117b
L. × chaytorae 104
L. × c. 'Richard Gray' 104
L. × c. 'Sawyers' 36, 104, 118i
L. lanata 104
lawns 13
leaves, collecting 16
Lenten rose see Helleborus × hybridus
Leucojum vernum 69b
Leucothoe 114
L. axillaris 'Curly Red' 102
L. fontanesiana 'Rainbow' 95, 114i
L. 'Girard's Rainbow' 114
L. LOVITA ('Zebonard') 102i, 114
L. SCARLETTA ('Zeblid') 102, 114, 130b
Leycesteria formosa 111i, 148b
Libertia peregrinans 101, 118i, 119
light, winter 134–5
Ligularia dentata 'Britt-Marie Crawford' 37
Ligustrum
L. lucidum 48b, 89i
L. ovalifolium 'Aureum' 97b
Lilium regale 37
Linnaeus 20
Liriope
L. muscari 'Variegata' 99b, 129
L. spicata 'Gin-ryu' ('Silver Dragon') 94
Lonicera
L. fragrantissima 143b
L. nitida 'Baggesen's Gold' 35–6, 96i, 99i, 131b
L. pileata 89b
L. × purpusii 37, 40, 85b
L. × p. 'Winter Beauty' 142i
Lophomyrtus × ralphii 'Kathryn' 102b
Loropetalum chinense f. rubrum 'Fire Dance' 102b
Luma apiculata 146
Lunaria annua 29i
L.a. var. albiflora 29
Luzula sylvatica 121
L.s. 'Aurea' 31, 97i, 121
L.s. 'Taggart's Cream' 121
Lychnis coronaria 129

M
Magnolia
M. grandiflora 89, 111i, 127b
M.g. 'Little Gem' 111
M. × soulangeana 111
M. stellata 32i, 111i
Mahonia 48
M. aquifolium 48, 102
M.a. 'Apollo' 48i, 82i, 122b
M.a. 'Atropurpurea' 48i, 102i, 119
M. japonica 102, 142i
M. lomariifolia 142

M. × *media* 122, 142
M. × *m.* 'Charity' 89b, 122i, 123, 142
M. × *m.* 'Lionel Fortescue' 142
M. × *m.* 'Underway' 142
M. × *m.* 'Winter Sun' 142i
M. pinnata 48
M. × *wagneri*
M. × *w.* 'Moseri' 48
M. × *w.* 'Pinnacle' 48i
Malus
M. bhutanica 27b
M. transitoria 26i
M. × *zumi* 'Golden Hornet' 26
marjoram see *Origanum vulgare*
Mertensia virginica 38
Meserve, Kathleen 106
Mexican orange blossom see *Choisya ternata*
Microbiota decussata 51, 101i
Milium effusum 'Aureum' 29, 125
mimosa see *Acacia dealbata*
Miscanthus 5i, 31, 150i, 151
M. sinensis 111b, 134i, 147i
M.s. 'Flamingo' 31
M.s. 'Gracillimus' 31
M.s. 'Variegatus' 31i
mistletoe see *Viscum album*
Molinia caerulea 31
M.c. 'Variegata' 31i, 147i
Myrtus communis 139i
M.c. subsp. *tarentina* 139

N
Nandina domestica 27b
N.d. 'Fire Power' 95
Narcissus
N. bulbocodium 69b, 111b
N. 'Cedric Morris' 66
N. cyclamineus 111b
N. 'February Gold' 33, 66
N. 'Rijnveld's Early Sensation' 66i
N. 'Tête-à-tête' 33, 66, 155i
N. 'Thalia' 36
Nerine bowdenii 78i
Nicotiana affinis 73

O
Ophiopogon planiscapus 'Nigrescens' 39, 56i, 65, 103i, 130i, 141i, 149i, 155i
Origanum vulgare 153
O.v. 'Aureum' 152i, 153
O.v. 'Polyphant' 61i, 152, 153
Osmanthus
O. armatus 89b
O. × *burkwoodii* 35, 89i
O. delavayi 89
O. heterophyllus 49
O.h. 'Goshiki' 49, 95i, 119b
O.h. 'Gulftide' 49
O.h. 'Variegatus' 49i, 92–3, 155i
Ozothamnus rosmarinifolius 'Silver Jubilee' 107b

P
Pachysandra terminalis 123i
P.t. 'Variegata' 123
Paeonia
P. broteroi 27
P. cambessedesii 27i
P. mlokosewitschii 32, 33i
parsley see *Petroselinum crispum*
perennials
flowering 72–9
herbaceous 17–18, 32–3
multicoloured foliage 95
persistent 55–61
for seedheads 28–9
variegated foliage 94–5
periwinkle see *Vinca*
Persicaria
P. affinis 101
P.a. 'Darjeeling Red' 101
P.a. 'Donald Lowndes' 101
P. microcephala 'Red Dragon' 54–5
P. milletii 38
Petasites
P. albus 120i
P. paradoxus 120
Petroselinum crispum 152–3
pheasant berry see *Leycesteria formosa*
Phillyrea
P. angustifolia 49b, 150–1
P. latifolia 151
Phlomis
P. fruticosa 105, 118i
P. italica 105i, 118
P. russeliana 28i, 29
P. tuberosa 'Amazone' 29i
Phoenix canariensis 53b
Phormium 52, 130
P. 'Bronze Baby' 52
P. cookianum 'Flamingo' 52
P. 'Dark Delight' 130
P. 'Maori Chief' 130
P. 'Maori Sunrise' 130
P. 'Platt's Black' 52, 103
P. 'Surfer Bronze' 101b, 130
P. tenax
P.t. 'Nanum Purpureum' 103
P.t. Purpureum Group 52
P.t. 'Variegatum' 52
P. 'Tom Thumb' 103
P. 'Yellow Wave' 52, 148i
Photinia
P. × *fraseri* 'Red Robin' 49b, 116i
P. 'Redstart' 116
Phyllostachys
P. aureosulcata f. *aureocaulis* 148b
P. nigra 151
P. vivax f. *aureocaulis* 151
Physocarpus opulifolius 'Dart's Gold' 36, 94i, 148i
Picea
P. glauca ALBERTA BLUE ('Haal') 106
P. orientalis 'Aurea' 51b
P. pungens 107
P.p. 'Globosa' 101, 115
P.p. 'Koster' 107i
Pieris 85, 114–15
P. japonica
P.j. 'Christmas Cheer' 85i, 114
P.j. 'Katsura' 115
P.j. 'Little Heath' 93i, 114i
P.j. 'Valley Valentine' 115b
P.j. 'White Rim' 93
Pileostegia viburnoides 125b
Pinus
P. bungeana 146
P. densiflora 'Alice Verkade' 51
P. mugo
P.m. 'Mops' 51b, 90
P.m. 'Ophir' 51
P.m. 'Winter Gold' 51i, 115
P. sylvestris 145b
Pittosporum 92
P. 'Garnettii' 36, 92i
P. tenuifolium 49b
P.t. 'Irene Paterson' 92i, 155i
P.t. 'Purpureum' 103b
P.t. 'Silver Queen' 92
P.t. 'Tom Thumb' 78i, 102i, 131b, 155i
P.t. 'Warnham Gold' 97i
planning 34–9
Platycladus orientalis 'Aurea Nana' 51, 97i
Pleioblastus auricomas 98i
Podocarpus 51
P. 'Chocolate Box' 51
P. nivalis 51
P.n. 'Bronze' 51i, 101b
P. 'Young Rusty' 51
Polygonatum 55, 125
Polygonum affine see *Persicaria affinis*
Polypodium
P. cambricum 91
P. vulgare 91i
Polystichum setiferum 91i
ponds 18
Populus tremula 'Pendula' 150
Portugal laurel see *Prunus lusitanica*
Postill, Alan 137
Potentilla fruticosa 'Elizabeth' 35
pots/containers 128–31
Primula
P. Cowichan series 129b
P. 'Gigha' 79
P. 'Guinevere' 38
P. vulgaris 33i, 81i, 123b
P.v. subsp. *sibthorpii* 79i
P. Wanda Group 83
Prostanthera cuneata 139i
pruning, winter 19
Prunus 80–1
P. incisa 81
P.i. 'February Pink' 81
P.i. 'Kojo-no-mai' 81
P.i. 'Mikinori' 81
P.i. 'Praecox' 81i
P. laurocerasus 88, 119
P.l. 'Castlewellan' 92i, 94i
P.l. ETNA ('Aubri') 88i
P.l. 'Otto Luyken' 88, 119i
P.l. 'Rotundifolia' 88
P. lusitanica 47i
P.l. 'Variegata' 47i
P. maackii 145b
P.m. 'Amber Beauty' 100i
P. mume 81
P.m. 'Alboplena' 81
P.m. 'Beni-chidori' 80i, 81, 143b
P.m. 'Pendula' 81
P. serrula 31, 135, 144i, 145b, 154i
P. × *subhirtella*
P. × *s.* 'Autumnalis' 80i–81
P. × *s.* 'Autumnalis Rosea' 81
Pseudowintera colorata 95i
Pulmonaria 79, 81i, 94i
P. 'Diana Clare' 53
P. 'Mary Mottram' 37
P. rubra 78i, 79i
P.r. var. *albocorollata* 79i
P.r. 'Barfield Ruby' 79
P. saccharata 111b
Purdom, William 143
purple moor grass see *Molinia caerulea*
Puschkinia scilloides 81
Pyracantha 26, 125–6
P. DART'S RED ('Interrada') 125–6
P. 'Golden Charmer' 27b
P. 'Harlequin' 126
P. 'Orange Glow' 26i, 117b, 125i
P. rogersiana 'Flava' 126
P. SAPHYR ORANGE ('Cadange') 12

Q
quince, ornamental see *Chaenomeles*

R
Ramsdale, Mary 59
Ranunculus ficaria 68
R.f. 'Brazen Hussy' 38, 68i, 101
Rhamnus alaternus 'Argenteovariegata' 93b, 125
Rheum
R. 'Ace of Hearts' 32
R. × *hybridum* (rhubarb) 153
Rhododendron 85, 113–14
bulbs, underplanting 113
R. 'Christmas Cheer' 113
R. dauricum 37i, 113i, 114
R.d. 'Midwinter' 113
R. 'Joanna' 103b
R. mucronulatum 113
R.m. 'Cornell Pink' 113i
R. 'Nobleanum Venustum' 113i
R. 'Olive' 114i
R. PJM Group 'Peter John Mezitt' 103b
R. ponticum 'Silver Edge' 93
R. 'Praecox' 85i, 115b, 139
Ribes
R. laurifolium 84, 110i
R. sanguineum 32i
R.s. WHITE ICICLE ('Ubric') 110i
Rodgersia pinnata
R.p. 'Buckland Beauty' 29
R.p. 'Maurice Mason' 29
Rosa
R. 'Felicia' 24–5, 39, 41i
R. 'Geranium' 27b
R. ICEBERG ('Korbin') 36
R. MOLINEUX ('Ausmol') 36
R. rugosa 25, 119
R.r. 'Alba' 119
R. virginiana 25i, 119i
R. 'Will Scarlet' 25
rosemary see *Rosmarinus officinalis*
Rosemoor, RHS Garden, Devon 11i
Rosmarinus officinalis 89, 117b, 152i, 153
R.o. 'Miss Jessopp's Upright' 153
rowan see *Sorbus aucuparia*
Royal Horticultural Society 5, 11i
Rubus
R. cockburnianus 147i, 149
R.c. 'Goldenvale' 141i, 149i
R. phoenicolasius 149b
R. thibetanus ('Silver Fern') 149
Ruscus aculeatus 47i, 123b
Ruta graveolens 107
R.g. 'Jackman's Blue' 107i

S
sacred bamboo see *Nandina domestica*
sage see *Salvia officinalis*
Salix 85, 111, 120, 148
pruning 19
S. acutifolia 'Blue Streak' 85
S. alba
S.a. 'Dart's Snake' 149b
S.a. var. *vitellina* 121, 148
S.a.v.v. 'Britzensis' 121i, 148
S. babylonica var. *pekinensis* 'Tortuosa' 15, 120–1i
S. caprea 32i
S. daphnoides 85, 149b
S.d. 'Aglaia' 85
S. 'Erythroflexuosa' 120i, 149b

S. gracilistyla 'Melanostachys' 85i
S. udensis 'Sekka' 85i
Salvia officinalis
S.o. 'Icterina' 153
S.o. 'Purpurascens'
S.o. 'Tricolor' 153
sandy soil 118–19
Santolina 118, 139i
S. chamaecyparissus 104, 118i
S. rosmarinifolia subsp. *rosmarinifolia* 104
Sarcococca 36, 122, 143
S. confusa 49b, 74, 89b, 115b, 131b, 143
S. hookeriana
S.h. var. *digyna* 38, 85b, 143i, 155i
S.h.v.d. 'Purple Stem' 143
S.h. var. *humilis* 123b
S. ruscifolia 122i, 143
Sasa veitchii 151i
Savill Garden, Surrey 11
Saxifraga
S. 'Aureopunctata' 99b
S. fortunei 91, 123
S. umbrosa 56
scented plants 136–43
Scilla
S. bifolia 69b
S. mischtschenkoana 69i
S. siberica 113i, 129b
sedges see grasses/sedges
Sedum 119
S. 'Herbstfreude' 22–3i
S. 'Purple Emperor' 119
S. spectabile 119i
S.s. 'Iceberg' 119
seedheads 28–9
Senecio 'Sunshine' see *Brachyglottis* 'Sunshine'
sensory gardens 132–53
light in 134–5
movement in 150–1
scented plants for 136–43
taste, plants for 152–3
texture in 144–9
shade
planting schemes for 38, 124–6
silver foliage for 106
woodland 122–3
Sheffield Winter Garden 11
shrubs
autumn-flowering 24
evergreen classics 44–9
for all-year interest 34–6
multicoloured foliage 95
pruning 19
variegated 92–4
winter-flowering 82–5
Sir Harold Hillier Gardens, Hampshire 10i, 18i, 141
Sisyrinchium striatum 53b
Skimmia 49
S. anquetilia 139
S. × *confusa* 139i
S. × *c.* 'Kew Green' 49i, 115b, 129–30
S. japonica
S.j. 'Fragrans' 49, 143b
S.j. 'Kew White' 49
S.j. 'Nymans' 49i
S.j. 'Rubella' 13, 33, 49i, 85b, 115, 129, 155i
snowberry see *Symphoricarpos albus*
snowdrop see *Galanthus*
snow gum see *Eucalyptus pauciflora* subsp. *niphophila*
soil
acid 112–14
chalky 116–17
clay 110–11
digging 16
pH of 114
sandy 118–19
wet/waterlogged 120–1
Solomon's seal see *Polygonatum*
Sophora SUN KING ('Hilsop') 127b
Sorbus
S. aucuparia 26
S. 'Joseph Rock' 26i
Spiraea japonica 'Goldflame' 32i
spotted laurel see *Aucuba*
spring 32–3
spurge laurel see *Daphne laureola*
Stachys
S. byzantina 106, 118i
S. officinalis 'Rosea Superba' 37
Stachyurus praecox 33i
stems, decorative 15, 147–9
Stipa 30
S. arundinacea see *Anemanthele lessoniana*
S. gigantea 30
S. tenuissima 30i, 39, 151, 155i
strawberry tree see *Arbutus unedo*
summer colour, planning for 35–7
sun, planting schemes for 39, 126–7
sweet bay see *Laurus nobilis*
sweet violet see *Viola odorata*
Symphoricarpos
S. albus 27
S. × *doorenbosii*
S. × *d.* 'Magic Berry' 27i
S. × *d.* 'Mother of Pearl' 27

T
Tanacetum parthenium 'Aureum' 97, 130
Taxus 50
T. baccata 90i
T.b. 'Fastigiata' 12, 50i, 90
T.b. 'Fastigiata Robusta' 50
T.b. 'Ivory Tower' 50
T.b. 'Standishii' 50i
Tellima 60, 61i
T. grandiflora
T.g. 'Forest Frost' 60i
T.g. Rubra Group 37, 39, 60, 86–7i, 100i, 102i, 155i
Teucrium fruticans 107b
texture, plants for 144–9
Thamnocalamus crassinodus 'Kew Beauty' 151
Threave Gardens, Dumfries and Galloway 11
Thuja 51, 139i
T. occidentalis
T.o. 'Ericoides' 101b
T.o. 'Rheingold' 51i, 96–7, 119b
T. plicata
T.p. 'Rogersii' 51
T.p. 'Stoneham Gold' 51
Thymus 139i
T. 'Doone Valley' 152, 153
T. serpyllum 90
T. vulgaris 153
T.v. 'Silver Posie' 152i, 153
Tiarella 60
T. cordifolia 'Glossy' 60
T. 'Skid's Variegated' 60i, 95
T. wherryi 'Bronze Beauty' 60
Tolmiea menziesii 'Taff's Gold' 99b
Trachelospermum
T. asiaticum 127i
T. jasminoides 127i
T.j. 'Variegatum' 127
Trachycarpus fortunei 53b
trees
bark, textured/patterned 144–6
pruning 19
winter-flowering 80–2
Tulipa turkestanica 69b

U
Uncinia rubra 31i, 101i, 111b, 155i

V
Vaccinium corymbosum 149b
variegated foliage 92–5, 98–9
reversion in 93
vegetables 153
Verbascum bombyciferum 106i
Verbena bonariensis 41
Veronica peduncularis 'Georgia Blue' 33i, 57
Viburnum
pruning 143
V. × *bodnantense* 40, 143
V. × *b.* 'Charles Lamont' 143i
V. × *b.* 'Dawn' 143i
V. × *b.* 'Deben' 143
V. × *burkwoodii*
V. × *b.* 'Anne Russell' 126i
V. davidii 48i, 84, 123b
V. farreri (*V. fragrans*) 15, 36, 37, 143
V.f. 'Candidissimum' 143b
V. × *globosum* 'Jermyns Globe' 35i
V. tinus 49b, 84i, 89b, 123b
V.t. 'Eve Price' 85, 131b
V.t. 'Gwenllian' 84i–85
V.t. 'Purpureum' 85
V.t. 'Variegatum' 85
views, from windows 12–13, 134
Vinca 98, 123
V. difformis 49, 123
V. major 49
V. minor 49
V.m. f. *alba* 123i
V.m. 'Atropurpurea' 49
V.m. 'Aureovariegata' 98
V.m. 'Azurea Flore Pleno' 49
V.m. 'Illumination' 49i, 98i, 103
vine weevil 59
Viola
V. cornuta 36, 37
V. labradorica 78
V. odorata 77i–78i
V.o. 'Alba' 78
V.o. 'King of Violets' 78
V.o. Rosea Group 78
V. riviniana 78
V.r. Purpurea Group 78–9
Viscum album 14

W
walls/fences, plants for 124–7
weeping aspen see *Populus tremula* 'Pendula'
wet/waterlogged soil 120–1
white forsythia see *Abeliophyllum distichum*
White Windows, Hampshire 6–7i, 61i, 90i, 154i
wild ginger see *Asarum europaeum*
willow see *Salix*
Wilson, Ernest 149
windows, views from 12–13, 134
winter aconite see *Eranthis hyemalis*
winter gardens
history of 8i–9
to visit 10–11
winter jasmine see *Jasminum nudiflorum*
winter protection 18
winter's bark see *Drimys winteri*
wintersweet see *Chimonanthus praecox*
Wisley, RHS Garden, Surrey 11i
witch hazel see *Hamamelis*
wood spurge see *Euphorbia amygdaloides* var. *robbiae*
woodland 122–3
wreaths 15

Y
Yucca 52
Y. filamentosa 'Variegata' 52
Y. flaccida 'Golden Sword' 52
Y. gloriosa 52
Y.g. 'Variegata' 52

ACKNOWLEDGMENTS

David & Charles would like to thank all members of the team that created this book. They have put so much effort and work into making it a fitting tribute to Jane's memory. We would especially like to thank Andy McIndoe for his unstinting hard work. Sue Gordon, Robin Whitecross and Lesley Riley were totally dedicated and worked far beyond the call of duty, as always.

OutHouse Publishing would like to thank Barry Sterndale-Bennett, Jennifer Harmer and Sue Ward for their assistance during the final stages of producing *The Winter Garden*.

PICTURE CREDITS

The publishers would like to acknowledge with thanks all those whose gardens are pictured in this book.

All photography by Jane Sterndale-Bennett, Andrew McIndoe and John Hillier except:
Pip Bensley 47 GC2; Adrian Bloom/Bloom Pictures 9a; Duncan and Davies 112b; Flora Pix 21b, 26d, 27a, 29b, 77b, 78 GC3, 93a, 94(2), 99a, 99c, 99d, 99e, 100c, 101a, 116b, 122b 124b, 137a, 137b, 139(5), 139(8), 140(4), 146d, 147(3), 147(4), 150a; Sue Gordon 15a, 119b; Kevin Hobbs 2c, 10a, 11c, 54a, 57b, 58c, 58a, 59b, 59c, 64a, 69d, 69e, 76c, 81(1), 81(4), 95(4), 113(2), 117e, 117f, 118e, 118f; The Big Grass Company 121d; New Leaf Plants 127(3); Clive Nichols 41a; Stephen Record 11b